Insect Pest Management Guidelines for California Landscape Ornamentals

C. S. Koehler, Entomologist

Cooperative Extension **University of California**
Division of Agriculture and Natural Resources
Publication 3317

WARNING ON THE USE OF CHEMICALS

Pesticides are poisonous. Always read and carefully follow all precautions and safety recommendations given on the container label. Store all chemicals in their original labeled containers in a locked cabinet or shed, away from food or feeds, and out of the reach of children, unauthorized persons, pets, and livestock.

Recommendations are based on the best information currently available, and treatments based on them should not leave residues exceeding the tolerance established for any particular chemical. Confine chemicals to the area being treated. THE GROWER IS LEGALLY RESPONSIBLE for residues on the grower's crops as well as for problems caused by drift from the grower's property to other properties or crops.

Consult your County Agricultural Commissioner for correct methods of disposing of leftover spray material and empty containers. **Never burn pesticide containers.**

PHYTOTOXICITY: Certain chemicals may cause plant injury if used at the wrong stage of plant development or when temperatures are too high. Injury may also result from excessive amounts or the wrong formulation or from mixing incompatible materials. Inert ingredients, such as wetters, spreaders, emulsifiers, diluents, and solvents, can cause plant injury. Since formulations are often changed by manufacturers, it is possible that plant injury may occur, even though no injury was noted in previous seasons.

Issued in furtherance of Cooperative Extension work, Acts of May 8 and June 30, 1914, in cooperation with the U.S. Department of Agriculture.
Kenneth R. Farrell, Director of Cooperative Extension, University of California.

3m-rep-1/88-LE/VG

Contents

The author is C. S. Koehler, Entomologist,
Cooperative Extension, University of California, Berkeley.

Foreword

This manual was developed to assist the arborist, public- and private-sector grounds maintenance person, nursery operator, landscape architect, consultant and other advisory person, and student of ornamental horticulture with the management of insect pests of California landscape ornamentals.

Several background sections on the theory and practice of pest management technology precede the heart of the manual—tables that enumerate the various pests of common landscape plants, with options, if they exist, for managing them. In some instances certain plants are not covered because insufficient information is available on the pests that attack them, or because methods of managing known pests have never been evaluated on those plants. It is anticipated that future revisions of this manual will include additional plants and their insect pests.

Acknowledgments

The author is grateful to many people for discussions and interaction with them over the years about insects associated with ornamentals, or for their help in the preparation or review of the manuscript. They are William W. Allen, A. D. Ali, Leslie W. Barclay, Barbara A. Barr, Pamela S. Bone, Ed Brennan, Leland R. Brown, Joseph H. Connell, Laurence R. Costello, John A. Davidson, James A. Downer, Gordon W. Frankie, Denice Froehlich, Mike Haines, W. Douglas Hamilton, Gary W. Hickman, Warren T. Johnson, Emile Labadie, Allen Lagarbo, Wayne S. Moore, William H. Olson, Edward J. Perry, Michael Raupp, Theodore W. Stamen, Pavel Svihra, and Janet S. Zalom.

The assistance of the Elvenia J. Slosson Endowment Fund for Ornamental Horticulture for its financial support in the preparation of this manual is gratefully acknowledged.

Nature of Insect Damage to Woody Ornamentals

Insects and their near and distant relatives (such as mites, slugs, and snails) seldom directly kill a woody plant on which they feed. There are exceptions, of course, but whenever plant death occurs, a cause other than insects should be considered first.

For the most part, insects are responsible for disfiguring ornamental plants in a variety of ways. Disfigurement may cause the tree or shrub to appear unsightly or to give less shade or visual screening. Sometimes the plant's rate of growth may be slowed, its shape or form changed, or it becomes predisposed to attack by disease-causing pathogens, or other insects. Disfigurement is sometimes overshadowed by the nuisance some insects create. For example, sticky honeydew collecting on automobiles, sidewalks, or lawn furniture may be the result of aphid or soft (unarmored) scale populations too low to attract attention to whatever damage they cause directly to the tree. Falling oak galls, caterpillar droppings, and insects crawling or pupating on exterior residential walls or window screens may be more significant than the actual plant damage caused by these same insects.

Diagnosing Plant Problems Caused by Insects

Most people recognize when a plant is unsightly, growing poorly, or causing a nuisance; however, determining the cause of poor plant performance can be difficult.

Insects and their relatives must feed to survive and reproduce. Most feeding pests cause visible and predictable changes in the plant's appearance, enabling the trained observer to narrow the possibilities as to the identity of the offending pest. Plant identification and the recognition of damage symptoms are the important first steps in the diagnostic process. Insect-caused damage symptoms can be grouped conveniently into five categories:

—Chewed or tattered foliage or blossoms
—Stippled (flecked), yellowed, bleached, or bronzed foliage
—Distortion of plant parts
—Dieback of plant parts
—Presence of insect (or insect-related) products

Category I Symptoms:

Chewed or Tattered Foliage or Blossoms

(See color plates on page 4 for Category I symptoms.)

The symptoms of chewed or tattered foliage or blossoms are, in most cases, caused by an insect or one of its relatives that has chewing mouthparts. The diagnostician can immediately dismiss from consideration a large group of pests with sucking or rasping mouthparts, for except in unusual situations, the feeding of such insects does not result in tattered plant parts. The most important pests causing chewed or tattered foliage and blossoms are:

—Larvae of moths and butterflies
—Larvae of, and sometimes adult, beetles
—Sawfly larvae
—Grasshoppers
—Snails and slugs

Keep in mind that several plant diseases often result in tattered or shot-holed leaves on *Prunus* spp.

Category II Symptoms: Stippled (Flecked), Yellowed, Bleached, or Bronzed Foliage

(See color plates on page 5 for Category II symptoms.)

When a plant's foliage is stippled, yellowed, bleached, or bronzed, and no loss in the physical integrity of foliage surface is seen, the injury is caused by insects (or their relatives) with some form of sucking—not chewing—mouthpart. These symptoms often begin with stippling, or flecking, of leaves, resulting from insertion of sucking mouthparts into the leaf. Chlorophyll is withdrawn or destroyed at the point of penetration. Tiny discolored (stippled) areas appear on affected foliage. With large numbers of attacking insects, these stippled areas coalesce, giving rise to leaves that appear partly or mostly bleached, yellowed, bronzed, or silvered.

Actually, not all arthropods that cause Category II symptoms possess true sucking mouthparts. Some have rasping or puncturing mouthparts and the pest imbibes plant fluids that leak from the ruptured plant cells. For convenience, however, insects with rasping or puncturing mouthparts are included with the true piercing-sucking type. Those arthropods typically responsible for causing Category II symptoms include:
—Spider mites
—Leafhoppers
—Plant bugs
—Lace bugs
—Thrips
—Aphids
—Psyllids

Category III Symptoms:

Distortion of Plant Parts

(See color plates on page 6 for Category III symptoms.)

Plant distortion may be curled or cupped leaves, twisted growing points, or galls (swellings or growths) of various types on leaves, flowers, twigs, or stems. In many cases the arthropod responsible may not be readily visible on the surface of the affected plant part. Pests responsible for Category III symptoms include:
—Aphids
—Thrips
—Cynipid (gall) wasps
—Larvae of certain moths
—Eriophyid (gall, blister, bud, or rust) mites

Category IV Symptoms:

Dieback of Plant Parts

(See color plates on page 7 for Category IV symptoms.)

Leaf, twig, or branch dieback characterize category IV symptoms. In a few cases death of the entire plant may occur. Twigs and branches of deciduous plants that die during the growing season often retain dead leaves well into the subsequent dormant season, for the leaf abscission layer will not have been formed before the plant part died. Parts of nondeciduous plants that die at any time will retain dead leaves for long periods. Whenever these symptoms are seen, the following arthropods are suspect:
—Scale insects
—Moth or beetle larvae that bore
—Cynipid (gall) wasps

Category V Symptoms:

Insect (or Insect-related) Products

(See color plates on pages 8-10 for Category V symptoms.)

Some insects produce evidence of their presence beyond that of plant injury. Many of these products remain intact for weeks—and often months—after the pest has completed its activities. The most commonly seen products, and the pests responsible for them, are:
—Honeydew and sooty mold: aphids, soft scales, leafhoppers, mealybugs, psyllids, and whiteflies
—Dark fecal specks: greenhouse thrips and lace bugs
—Tents, webs, silken mats: tent caterpillars, webworms, and leafrollers
—Spittle: spittlebugs
—Cast skins: aphids, leafhoppers, and lace bugs
—Pitch masses: larvae of certain moths, such as sequoia pitch moth
—Pitch tubes: certain scolytid bark beetles
—Flocculence (cottony waxy material): adelgids, mealybugs, certain scales, and aphids
—Slime: snails and slugs

These five symptom categories do not involve five completely different groups of insects or their relatives, for a single kind of pest may cause more than one kind of symptom. Aphids, for example, cause symptoms of yellowing (Category II) and plant distortion (Category III), and are responsible for products such as honeydew, sooty mold, cast skins, and flocculence (Category V). Some species of aphids commonly cause several kinds of symptoms concurrently. Scale insects may cause dieback of plant parts (Category IV) and leave products on plants (Category V). Snails and slugs cause tattered foliage (Category I) and leave slime trails (Category V). Keep in mind that various agents, such as plant diseases, herbicides, or cultural problems, may cause symptoms similar to those caused by insects and their relatives.

As a first step in the problem diagnosis process, let the *plant* serve as an early indicator of the trouble. The next step is to search for pests capable of causing the symptoms. Keep in mind that the pests found must occur in numbers sufficient to cause the damage noted, if the diagnosis is to be accurate. Remember, too, that more that one kind of pest may be present on the plant, giving rise to multiple symptoms.

Table 7, beginning on page 20, is organized so that the damage symptoms are described in the first column of every page to emphasize that it is the change in the plant's appearance from the "normal" appearance that first suggests poor plant performance to the arborist, landscape manager, or homeowner. For pests that are not covered in the tables, the diagnostic scheme described above should help the diagnostician to identify the type of pest responsible for the symptoms.

Color Plates

Category I Symptoms	— Chewed or tattered foliage or blossoms
Category II Symptoms	— Stippled (flecked), yellowed, bleached, or bronzed foliage
Category III Symptoms	— Distortion of plant parts
Category IV Symptoms	— Dieback of plant parts
Category V Symptoms	— Presence of insect (or insect-related) products

Leaf of Mahonia (Oregon grape) injured by the barberry looper.

Skeletonization of elm leaves by the larva of the elm leaf beetle.

Notching, or scalloping, of leaves of toyon by an *Otiorhynchus* weevil adult.

Pyracantha leaves bronzed by spider mite feeding.

Photonia leaf bleached by the greenhouse thrips (*upper*), as compared to normal leaf (*bottom*).

Monterey pines bleached by spider mite feeding.

Category III Symptoms

Distortion of toyon leaves caused by the toyon thrips.

Leaf galls caused by the gallwasp ***Andricus kingi*** on valley oak.

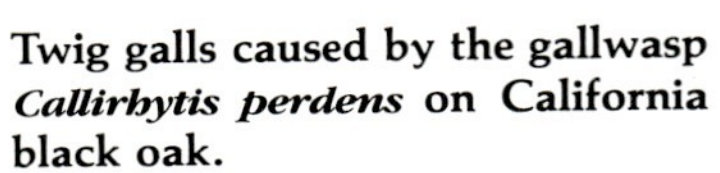

Twig galls caused by the gallwasp ***Callirhytis perdens*** on California black oak.

Fuchsia growth distorted by the fuchsia gall mite.

Dieback of white alder caused by the larva of the flatheaded borer *Agrilus burkei.*

Dieback of portions of live oak leaves caused by the two-horned oak gallmaker.

Twig dieback in valley oak caused by the oak pit scale.

Fecal specks of the greenhouse thrips on toyon leaves.

Spittlebug masses on rosemary.

A pitch tube made by the red turpentine beetle.

continued

Sooty mold on California bay laurel, resulting from honeydew produced by the laurel aphid.

Fall webworm tents in Lombardy poplar.

continued

Woolly apple aphid flocculence on pyracantha.

Egg sacs of an iceplant scale on iceplant.

When Should Control Action Be Taken?

One of the principles of modern pest management is that a pest must be present—or threatening to be present—in numbers sufficient to cause unacceptable or intolerable damage, before action is taken to control it. Damage refers to the array of insect-caused disfigurement of plants and their surroundings, described earlier. The above principle has its origin in agricultural insect pest management, and at present is only partially applicable to ornamentals pest management.

The actual pest level at which control action should be taken in landscape plantings is an elusive concept, and depends on:

—The location of the plant in the landscape. One in a back yard, along a freeway, or in a seldom-used portion of a park, typically is less visible than one in a front yard or other conspicuous setting. Ordinarily, higher levels of pest damage can be tolerated on the less-visible plants. Yet individual people differ in their perception of "unacceptable" or "intolerable" damage, regardless of where the plant is situated.

—The species of pest involved. One likely to threaten the life of the plant (e.g., some scales) should be managed at lower levels than a pest species that transiently disfigures the plant. Similarly, some life-threatening species (e.g., certain bark beetles and other borers) that cannot be controlled once they have tunneled into the plant, are often justifiably managed by preventive means, without first having good information on the number of insects present.

—Whether the number of pests is increasing or decreasing. Regular monitoring of vegetation is the only way an accurate count can be determined. An increasing pest population may require control action if it has the propensity to reach the level at which unacceptable or intolerable damage is caused. A decreasing population may not require action, for whatever damage the pest is capable of causing has probably already occurred.

—The time of year the infestation occurs. An infestation occurring on the foliage of a deciduous tree in the fall, when leaf drop is imminent, ordinarily requires no control regardless of the level of pests present. Infestations developing early in the growing season, however, warrant close attention.

Sampling and Monitoring Landscape Pests and Their Natural Enemies

Various methods have been devised to sample, or estimate the numbers of, arthropods on trees and shrubs. Many of these procedures were first developed for use in agricultural crops, but several have resulted from investigations on landscape ornamental pests. The objectives of sampling or monitoring are to detect the presence, absence, or abundance of pests and their natural enemies, and to follow the progress of an arthropod population through time, by regular, periodic sampling. The goal of monitoring is to reach a decision as to whether, or when, a pest population requires control action. All sampling procedures share certain characteristics:

—They use a common sampling unit, such as leaves, terminals, beats, or minutes.

—The unit chosen must be consistent with the feeding habits of the pest population under observation. One would not select leaves as the sampling unit for scale insects that occur predominantly on twigs. Similarly, one would not count aphids on the younger leaves if they occur mostly on the older leaves.

—The number of samples taken must be adequate. What is "adequate" must be determined on a case-by-case basis, and by time and equipment constraints. Pests are seldom distributed uniformly over a tree or shrub. Similarly, every tree in a group of trees of that species will not be infested to the same degree. Generally, the number of samples taken from each plant at each interval is held constant over the entire sampling period, and over the entire group of plant samples.

—The sampling procedure must be standardized. It helps if the same person does all the samplings. If two or more persons are involved, they should check one another in a preliminary sampling exercise to determine that their sampling methods are the same.

—Written records of arthropod counts are kept by date, location, and person sampling, with a brief description of procedures used.

Because some insects and mites are quite small, the person sampling or monitoring pests and their natural enemies should carry and use a 10X hand lens.

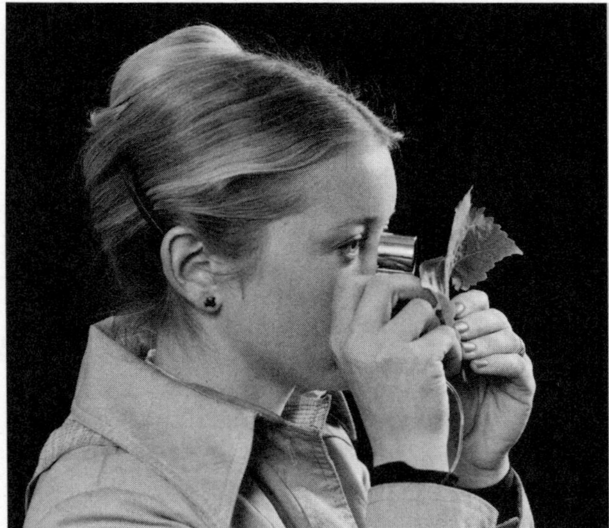

Hold the hand lens close to your eye and move in the specimen being examined until it is in focus.

Methods of Sampling

Counting insects on plant parts. Each sample is pruned or pinched from the plant and the number of arthropods present is counted immediately without magnification or under magnification, using a hand lens. The number of samples taken from each plant usually ranges from 5 to 25.

Variations include estimating pest numbers in 10s or 100s, when counts are high, or taking more samples (around 100) and recording only whether pests are present or absent on each unit.

Counting insects on plant parts is effective for sampling aphids, spider mites, some psyllids, and other arthropods that do not fly readily or drop from the plant when the sampling unit is removed. Sometimes only the immature insect stages are counted if the adults fly readily when the sample is taken.

Time counts. The person responsible for sampling counts the number of insects seen during a 1- or 2-minute visual search of the plant. Several such timed searches are made in different parts of the same plant (if large). This procedure is useful for large insects such as caterpillars, or for egg masses of insects on tree trunks or limbs. The plant is not damaged, and the insects counted are available for re-counting at the next scheduled sampling. Because it is difficult to count insects and keep track of time simultaneously, two persons are required for best results, unless an electronic alarm watch can replace the second person. The time-count is not a useful sampling method if the insect population is so high that the number of individuals tallied is limited by the rapidity by which each pest can be seen and counted.

Beating samples. A sampling tray is held horizontally just beneath plant foliage, and the foliage above is struck sharply a standard number of times (2 to 5) with a short stick, or with the other hand. Arthropods falling to the tray are immediately counted and then shaken off. This process is repeated several times around the periphery of the plant. An attempt is made to standardize the density of foliage beaten. The tray may be 1 square foot in surface area, or as small as a 5- or 6-inch circle (pads of paper or plastic disposable pie plates have often been used). The trap surface is usually white, to contrast with the insects being counted. This procedure has been used to sample such pests as psyllids, certain aphids, plant bugs, and spider mites.

Fecal pellet collections. Lepidopterous larvae, in particular, produce relatively large, solid, dark, fecal pellets, most of which fall to the ground beneath the plant. Using three to five shallow pans, paper cups, or sticky cards deployed beneath the foliage of infested trees, counts of pellets give an estimate of the larval population in the tree. The size of the individual pellets indicates whether the caterpillars are young or more mature. Collection traps are usually deployed for a 24-hour period each week. Traps are put out at times when no rain or sprinkler irrigation is expected in the next 24 hours.

Some tent-making and leaf-rolling caterpillars tend to deposit their fecal pellets such that few fall to the ground. The fecal trap method is not useful for sampling the number of those insects.

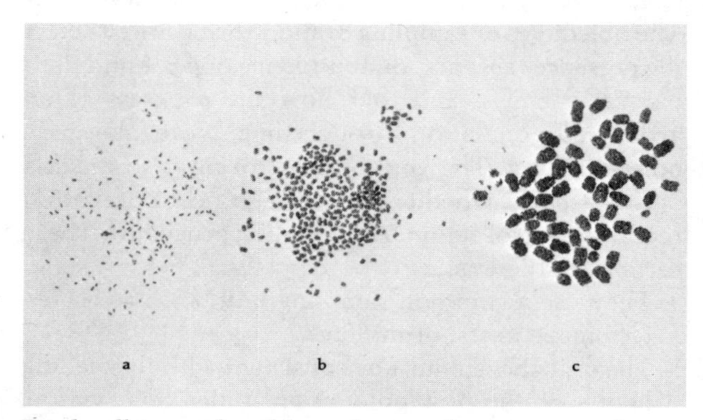

Fecal pellets produced by oakworm larvae: young larvae (*a*); midsized larvae (*b*); mature larvae (*c*).

Attractant traps. Devices containing synthetic or natural attractants, and that physically trap the insects attracted to the device, are useful for sampling several moth and beetle pests of ornamentals. They will trap only the motile adult stage, and in the case of most sex pheromone traps, only the male of the species.

Sampling and Monitoring Natural Enemies

When sampling for pests, the person sampling should also look for:

—Predators, such as ladybird beetle adults and larvae, syrphid fly larvae, lacewing larvae, and spiders.

—Evidence of parasitism, such as aphid "mummies," darkened greenhouse whitefly pupae, and scale insects with exit holes of the parasites.

—Signs of insect diseases, such as blackened dead caterpillars and dead, discolored aphids infected with fungi.

One can sometimes estimate the impact of biological control by counting the number of natural enemies per sampling unit, then calculating the ratio of affected pests to healthy ones. If signs of biological control are apparent, delay any insecticide treatment being considered. Sample again in a week for pests and natural enemies. If the natural enemy population is increasing faster than the pest population, consider no insecticide treatment, for natural enemies will often control the pest. If natural enemies are still active but the ratio has increased in favor of the pest, and pest numbers or plant unsightliness are approaching unacceptable levels, then consider the following measures when an insecticide is applied:

—If possible, use a material that is selective to preserve natural enemies (e.g., *Bacillus thuringiensis* for caterpillars), or a material that is minimally disruptive to natural enemies (e.g., insecticidal soap, horticultural oil, or systemic insecticides).

—If the use of selective or minimally disruptive materials is not possible, treat only the plants that are in immediate need of an application, leaving the untreated plants to serve as a reservoir for natural enemies.

—Various honeydew-producing insects, especially soft scales and aphids, are protected, or guarded, by ants, whose only interest is the honeydew food source. Ants interfere with maximum parasite and predator activity. The presence of ants on plants infested by honeydew-producing insects should therefore suggest control of the ants by insecticidal treatment of the base of the plant only. This is another kind of selective use of an insecticide. Sticky bands may be used instead of insecticide treatment.

Methods of Managing Pests

Four principal control options for existent or threatening populations of pests of ornamental plants are: biological control, cultural control, physical (or mechanical) control, and insecticidal control.

Biological Control

Every insect and mite pest has its complement of natural enemies (parasites and predators) that serve to reduce populations of the pest. Most natural enemies of landscape pests are native species, but others have been deliberately introduced into California, usually from foreign countries, and then released into the environment for biological control purposes. Biological control, then, is the use of parasites and predators for control of pest species.

Except in a few situations, the arborist or other landscape manager is seldom in a position to introduce new parasites and predators into the environment in which he or she works. The greatest opportunity to take advantage of biological control is to preserve those organisms that are already there. This is best accomplished if you:

—Learn to recognize important natural enemies of pests commonly encountered.

—Avoid the use of insecticides when a cultural or physical control measure would reduce pest numbers to an acceptable level.

—When a decision is made to use a pesticide, use one that has a minimal effect on parasites or predators (see above).

Cultural Control

Become informed about pest-resistant plants by reading research reports on the subject. The relative pest susceptibility of the groups of plants listed in tables 1, 2, 3, 4, and 5 on pages 14 to 16 has been established through experimentation. The plants listed first in each table are more highly resistant (and should be considered first for planting) than those listed later.

Plant growth-related practices useful in preventing or mitigating pest damage include: thinning rank vegetation, maintaining high plant vigor, judicious pruning, and planting resistant species or cultivars.

Thin rank vegetation to create a physical environment less favorable to certain soft scale insects. Observations and limited experience indicate that populations of black scale, *Saissetia oleae*, will decline if greater air circulation and more sunlight are allowed to enter the interior parts of dense plantings of oleander. Part of the scale decline may be due to improved accessibility of natural enemies to the scales. It is probable that other soft scales are similarly affected, and that vegetation control applies also to plants other than oleander.

Maintain high plant vigor to reduce losses from many boring insects. Bark beetles (family Scolytidae) of coniferous trees often attack trees in stress or in a reduced state of vigor caused by drought, root disease, root compaction, root damage done during construction, air pollution, a poor growing site, or even old age. Several bark beetles of non-coniferous trees, such as the smaller European elm bark beetle, *Scolytus multistriatus*, in elm; the shothole borer, *Scolytus rugulosus*, in plants of the genus *Prunus* spp.; and oak bark beetles, *Pseudopityophthorus* spp., in coast live oak, similarly appear to preferentially attack weakened trees. Many flatheaded borers (family Buprestidae) and roundheaded borers (family Cerambycidae) prefer to attack weakened, as compared to apparently vigorous, trees. Because control of most boring insects after they have infested plants is essentially impossible, preventing their attacks by cultural practices that contribute to high plant vigor makes good sense.

In a few documented instances, however, high plant vigor actually has contributed to high pest levels. Most of these instances involve sap-sucking pests. For example, high populations of the oleander aphid, *Aphis nerii*, can be prevented by reduced watering and pruning of plants. These practices discourage rank flushes of new growth preferred by the aphids. Yet, control of sucking pests is infinitely easier than control of established borers, and a strong case remains for the general principle of maintaining plants in good vigor.

Prune trees judiciously, and at the proper time of year, to reduce attacks by certain boring insects. Conifers, in particular, are susceptible to attack by several species of borers after tree trunks have been injured. This has been demonstrated for the sequoia pitch moth, *Synanthedon sequoiae*, and the red turpentine

Table 1. Cupressaceae resistant to the cypress tip miner, *Argyresthia cupressella*, ranked within genera from the most to the least resistant

Plant species	Number of insects per 100 grams foliage
Juniperus chinensis L. var. *sargentii* 'Glauca'	0.5
J. scopulorum Sarg. 'Erecta Glauca'	0.9
J. chinensis L. 'Kaizuka'	2.2
J. sabina L. 'Tamariscifolia'	5.1
J. virginiana L. 'Prostrata'	5.4
J. sabina L. 'Arcadia'	5.9
J. chinensis L. 'Pfitzerana Aurea'	7.6
J. virginiana L. 'Cupressifolia'	13.2
J. chinensis L. 'Robust Green'	18.4
J. chinensis L. 'Pfitzerana'	21.0
Thuja plicata J. Donn. ex. D. Don.	0.7
T. occidentalis L.	38.9
Chamaecyparis lawsoniana (A. Murr.) Parl. 'Allumii'	18.9

Source: C. S. Koehler and W. S. Moore, "Resistance of Several Members of the Cupressaceae to the Cypress Tip Miner, *Argyresthia cupressella*." (*Journal of Environmental Horticulture* 1, no. 4, 1983:87-88.)

Table 2. Acacias resistant to the acacia psyllid, *Psylla uncatoides*, ranked from the most to the least resistant

Acacia species	Number of nymphs per plant tip
aspera	0
podalyriifolia	0
baileyana	.01
parvissima	.01
craspedocarpa	.02
armata	.06
karoo	.06
cariophylla	.10
giraffae	.10
dealbata	.18
gerardii	.18
albida	.23
collettiodes	.30
cultiformis	.51
decurrens	.51
robusta	.58
mearnsii	.64
cunninghami	.81
iteaphylla	.83
cyanophylla	1.02
triptera	1.14
saligna	2.26
obtusa	2.34
spectobilis	3.03
pendula	5.87
implexa	7.83
cyclops	10.60
longifolia	11.73
penninervis	12.09
melanoxylon	24.42
retinodes	30.11

Source: C. S. Koehler, W. S. Moore, and B. Coate, "Resistance of Acacia to the Acacia Psyllid, *Psylla uncatoides*." (*Journal of Environmental Horticulture* 1, no. 3, 1983:65-67.)

beetle, *Dendroctonus valens*, in Monterey pine; and for cypress bark beetles, *Phloeosinus* spp., and the cypress bark moth, *Laspeyresia cupressana*, in Monterey cypress.

To prune judiciously, avoid excessive pruning—including topping—which results in poor balance between root and top growth, and avoid making flush pruning cuts. Conifers have often died from bark beetle attack soon after being excessively pruned.

For several boring insects that have been studied in detail, pruning (or otherwise opening wounds in tree trunks) during the late winter through spring and summer period has resulted in far more severe borer infestations than when pruning was done during the fall or early winter months. Adult borers are inactive in the fall and winter, and pruning at that time allows the tree to begin the wound closure process before adult activity resumes the following spring or summer.

Table 3. Fuchsias resistant to the fuchsia gall mite, *Aculops fuchsiae*, ranked from the most to the least resistant

Fuchsia **species or cultivar**	**Gall mite injury rating**
Baby Chang	1.0
Chance Encounter	1.0
Cinnabarina	1.0
F. minutiflora	1.0
F. thymifolia	1.0
Isis	1.0
Mendocino Mini	1.0
Miniature Jewels	1.0
Ocean Mist	1.0
Voodoo	1.5
Golden West	1.7
Englander	1.8
F. arborescens	2.0
Dollar Princess	2.2
F. procumbens	2.3
Lena	2.3
Pink Marshmallow	2.3
Psychedelic	2.3
Golden Anne	2.5
Kaleidoscope	2.8
Raspberry	2.8
Tinker Bell	2.8
Troubadour	2.8
Angel's Flight	3.0
Bicentennial	3.0
Capri	3.0
Dark Eyes	3.0
First Love	3.0
Lisa	3.0
Louise Emershaw	3.0
Novella	3.0
Swingtime	3.0
Vienna Waltz	3.0
Kathy Louise	3.3
Marinka	3.3
Papoose	3.3
Jingle Bells	3.5
South Gate	3.5
F. magellanica	3.7
China Doll	3.8
Christy	3.8
Display	4.0

Source: C. S. Koehler, W. W. Allen, and L. R. Costello, "Fuchsia Gall Mite Management." (*California Agriculture*, 39, no. 7 and 8, 1985:10-12.)

Table 4. Pines resistant to the Nantucket pine tip moth, *Rhyacionia frustrana*, ranked from the most to the least resistant

Pinus **species**	**Percent tips infested**
armandii	0
attenuata	0
bungeana	0
canariensis	0
caribaea	0
coulteri	0
edulus	0
gerardiana	0
monophylla	0
montezumae	0
mugo	0
nigra	0
palustris	0
pinaster	0
pinea	0
thunbergiana	0
torreyana	0
flexilis	3
ponderosa	7
rigida	10
jeffreyi	10
oocarpa	13
halepensis	17
taeda	23
brutia	30
cembroides	33
roxburghii	33
sylvestris	33
sabiniana	33
patula	37
pseudostrobus	40
muricata	40
resinosa	53
contorta	56
radiata	76
echinata	80
glabra	83
densiflora	83
insularis	83
virginiana	86

Source: G. T. Scriven and R. F. Luck, "Susceptibility of Pines to Attack by the Nantucket Pine Tip Moth in Southern California." (*Journal of Economic Entomology*, 73, no. 2, 1980:318-320.)

The application of wound dressing compounds, in the belief that they will prevent insect attack, is not a good alternative to judicious pruning at the proper time and in the proper manner.

Plant trees and shrubs that are resistant to insect attack. Pests do not attack equally every species and cultivar of all plants; therefore, selecting and planting those least susceptible to pests—so long as they meet basic requirements of size, color, and form—greatly reduces or negates the need for chemical control of insect pests. The term resistant does not necessarily mean immune. Low numbers of pests often occur on resistant plants, but these levels would be considered insignificant under most circumstances.

Despite the huge benefits from using pest-resistant plants, this cultural practice has the following shortcomings and limitations:

—A plant that is resistant to an important pest today probably would not be resistant to a new pest entering the area at some later time.

Table 5. Rhododendrons resistant to adult root weevils (*Sciopithes obscures, Otiorhynchus sulcatus, O. singularis, Nemocestes incomptus,* and *Dyslobus* spp.) ranked from the most to the least resistant

Rhododendron hybrid	Adult root weevil injury rating
P. J. Mezzitt (P.J.M.)	100
Jock	92
Sapphire	90
Rose Elf	89
Cilpimense	88
Lucky Strike	83
Exbury Naomi	81
Virginia Richards	81
Cowslip	80
Luscombei	80
Vanessa	80
Oceanlake	80
Dora Amateis	79
Crest	79
Rainbow	76
Point Defiance	76
Naomi	76
Pilgrim	76
Letty Edwards	76
Odee Wright	76
Moonstone	73
Lady Clementine Mitford	72
Candi	72
Graf Zeppelin	71
Snow Lady	71
Loderi Pink Diamond	71
Faggetter's Favourite	70

Source: A. L. Antonelli and R. L. Campbell, "Root Weevil Control on Rhododendrons." (Washington State University Cooperative Extension Bulletin 0970:1984.)

—Eventually, some insect pests have shown an ability to adapt to resistant plants. These new strains or "races" of pests are called biotypes.

—If a little-used plant should become widely used because it has been found resistant to an important pest, or for any other reason, the chances are great that an increasing number of insects would become pests of that plant. The thornless honeylocust cultivars, when first developed, were advertised as pest-resistant. If they ever were, they are no longer.

—A plant resistant to an important arthropod pest may be susceptible to serious plant diseases, or have other problems.

—If the landscape is already established, pest-resistant plants are of no use until new trees or shrubs are required.

Become informed about pest-resistant and other low maintenance plants by consulting with colleagues in the area. You will learn, for example, that:

—Coast redwood, *Sequoia sempervirens,* is far less pest-prone than Monterey pine.

—Tam (or Spanish) juniper, *Juniperus sabina* 'Tamariscifolia,' is more highly susceptible to the juniper twig girdler, *Periploca nigra,* than are other prostrate junipers.

—The maidenhair tree, *Ginkgo biloba,* is as pest-free a tree as one will encounter in most parts of California. Only the male tree is recommended for planting, since the fruit produced by the female tree is a nuisance.

Physical (or Mechanical) Control

In some cases a pest infestation may be controlled or averted by physical means. Examples follow.

Pruning off the tips of branches harboring colonies of caterpillars. Pruning branch tips when the larvae are young produces the best results. In the case of tent-making caterpillars, *Malacosoma* spp. (note, not all species of tent caterpillars make silken webs), the insects forage for foliage away from the tent on clear, warm days, but tend to return to the silken web before nightfall, and to remain there during cool, rainy weather. The branch tip containing the caterpillar colony, therefore, is best pruned off during inclement weather.

Larvae of the fall webworm, *Hyphantria cunea,* however, feed from within the protection of their tent and build an increasingly larger tent as more food is required for the colony. Their tents and colonies can be pruned from trees during any kind of weather, but results are best when the tents are small.

Several species of caterpillars that do not form tents tend to feed gregariously, especially when young. The redhumped caterpillar, *Schizura concinna*, and the spiny elm caterpillar (or larva of the mourningcloak butterfly), *Nymphalis antiopa*, feed together, and the entire colony can be removed by pruning off the branch terminal on which it is feeding.

Control by pruning, however, is of limited use on tall trees, or when infestations are large.

Reduce infestations of certain boring insects in living trees. The lead cable borer, *Scobicia declivis*, and branch and twig borer, *Polycaon confertus*, breed in dead oak, maple, bay, eucalyptus, and other hardwoods. When new adults emerge they frequently attack nearby living trees. Various scolytid bark beetles breed in stressed, dying, or recently dead wood of conifers. Removing such wood from the vicinity of living conifers reduces the likelihood of living trees becoming infested. This action should be taken whenever, and as soon as, potential breeding wood is detected, whether it is infested or not. The smaller European elm bark beetle, *Scolytus multistriatus* (vector of the Dutch elm disease fungus), and shothole borer, *Scolytus rugulosus*, similarly breed in declining or recently dead wood of elm and *Prunus* spp., respectively. Getting rid of such wood is recommended as a physical means of protecting nearby susceptible living trees. Burning, chipping, or burying dead trees to a depth of 18 inches or more, or physical removal of the tree to a distance of at least ¼ mile are effective means of disposal.

Prompt removal of the bark of freshly cut wood, a difficult task, will effectively negate its use as a breeding place for just about all insects, and will safely allow the wood to be kept as firewood. The bark that is removed need not be destroyed, for boring insects will not breed in it. Covering freshly cut wood (without first de-barking it) completely with a sound sheet of clear ultra violet ray-resistant plastic negates the wood being used as a breeding place for boring insects. Spraying freshly cut wood with an insecticide, if such wood will later be used as firewood, is not recommended.

Insecticidal Control

When threatened by a serious pest infestation, the landscape manager often has no choice but to apply an insecticide. Most are applied as sprays, but a few are used as liquid soil drenches or trunk injections.

Spray application methods. Sprays may be applied either by hydraulic or mist blower machinery. With the hydraulic sprayer, the insecticide is diluted in water and delivered to the target tree or shrub by high pressure and volume through a hose and gun. Mist blowers deliver a more highly concentrated pesticide by means of a high volume, high velocity air stream. That is, the pesticide is diluted largely in the air rather than in water. Mist blowing is most effective when there is no wind. Large mist blowers can treat tall trees rapidly, thoroughly, and with practically no runoff and little drift if there is no wind. Backpack models are often used on smaller trees and shrubs. Generally, emulsifiable concentrate formulations, if available, should be chosen for use in mist blowers because the abrasive characteristics of wettable powders eventually cause mechanical problems.

Systemic insecticides. When systemic insecticides are applied to the roots, leaves, or bark, or when injected into the vascular system, they are absorbed and translocated upward to kill insects feeding primarily on the foliage. When applied as sprays, they also kill insects by contact action. Acephate, dicrotophos, and dimethoate are systemics mentioned in this manual. All are organophosphate insecticides. Advantages of systemics are: relatively long residual life; protection of newly expanding foliage not present at the time of application; protection of plant parts such as growing points, which are difficult to physically penetrate with sprays; and reduced kill of beneficial insects.

Systemics will not kill all insects living in or on plants, however. Regardless of how the chemical is applied, it moves rather quickly to, or remains in, the foliage. Many sucking pests, such as aphids and thrips, and certain insects that chew foliage, can be easily controlled with the proper systemics. Systemics are not effective against most wood borers or bark-feeding scale insects, unless the proper material is applied as a spray that results in kill by contact action.

Trunk injection of systemic insecticides. The Mauget and Acecap systems are commercially available for injection and implantation, respectively, of systemics into the trunks of specific ornamental trees for insect control. Treatments are normally made in the growing season. To use both systems, first drill small holes into the tree at intervals around the circumference of the lower trunk. In the Mauget system, feeder tubes are inserted into each hole, then a plastic unit containing a prefilled small amount of insecticide is tapped onto the free end of each feeder tube, breaking a seal that allows the liquid to enter the feeder tube and then the tree's vascular system. After the plastic units have emptied, they and the feeder tubes are removed. In the Acecap system the entire small plastic unit containing a dry insecticide product is placed into each hole, seated just beneath the inner surface of the bark, and left there. Sap within the tree dissolves the insecticide and translocates it upward in the tree.

Research experiences in California have been more extensive with the Mauget, than with the Acecap, system. Elm leaf beetle control has been outstanding, and control of leaf feeding aphid species has been adequate to good, with Mauget products. Scales that feed on the bark of twigs or branches are not dramatically reduced in number. Researchers at the University of California have been unable to confirm the effectiveness of Mauget insecticides against the California fivespined ips in Monterey pine. Scientists doubt that control of boring insects can be practically achieved by using any type of injected or implanted insecticide treatment currently available.

Insecticide injection has much to recommend it if trees accept the systemic quickly, and if the insecticide is effective against pests for which it was intended. The equipment required is simple and inexpensive, trees can be injected during weather unsuitable for spraying, insecticidal drift is avoided, and negative impact on beneficial insects is believed minimal. Yet the possible chronic effects resulting from holes made in trees, which are required during the injection process, is not completely resolved, particularly for trees repeatedly treated.

Proper uses of pesticides. For a pesticide application to be a legal one, instructions on the container label must be followed. However, recent changes in federal and state laws have given applicators certain freedoms they did not enjoy earlier.

In 1983, the California Department of Food and Agriculture published its definitions of "conflict with labeling." Those that apply to landscape use follow. In California it is *not* a violation of label instructions to:

—Use less than the dosage specified on the label. For example, if the label indicates that 1 quart of the product should be mixed with every 100 gallons of water, and the finished spray applied to trees to the point of complete coverage, you may add less than 1 quart to each 100 gallons of water. Of course, the applicator should have good information that a reduced dosage will be effective.

—Use a concentration less than that specified on the label. Referring to the example above, you can use 1 quart of the particular product in 200, 300, or more gallons of water, and apply the finished spray, thereby reducing the concentration and its dosage rate.

—Apply a pesticide less frequently than specified on the label. If the label specifies that the application should be followed by a second application 10 days later, you may wait 12, 20, or more days before repeating the application. Or, you may choose never to repeat the application. Again, you should have good information that extending the interval between applications will not compromise effective pest control. You cannot apply a pesticide more frequently than specified on the label.

—Apply a pesticide to control a pest that is not named on the label, unless this is expressly prohibited by the labeling. However, the site to be treated (for example, the kind of tree) must be named—or not excluded—on the label. Sites given on some pesticide labels are quite specific, for example, azaleas, oaks, pine; whereas other labels give generic sites, for example, ornamentals, or shade trees.

—Use a method of application other than that specified on the label, so long as this other method is not prohibited by the labeling. All other label directions must be followed.

—Mix the pesticide with another pesticide or with a fertilizer, unless the label specifically prohibits such mixtures. Incompatibility of components in the mixture or plant injury (phytotoxicity) may result from mixing improper pesticides together, or mixing pesticides with fertilizers. The applicator should refer to a spray compatibility chart before mixing pesticides together. These charts are available from many agricultural chemical suppliers, or shown in the *Farm Chemicals Handbook*. When a mixture is to be made, the chances of phytotoxicity will be reduced if two pesticides of the same formulation (for example, wettable powders) are mixed together, rather than different formulations (for example, a wettable powder and an emulsifiable concentrate).

All labels for the same generic pesticide are not identical. As a result, dosages to be used may vary because of differences in strength between products containing the same pesticide. Also, sites that legally may be treated may differ from one label to another. For example, the label for the Ciba-Geigy product D.z.n. diazinon AG 500, an emulsifiable concentrate, lists specific ornamental plants to which that product may legally be applied: arborvitae, azalea, birch, and so forth. It would be illegal to apply that product to cypress, Douglas fir, or sycamore, for those sites (plants) are not listed on the label. However, the same manufacturer's product D.z.n. diazinon 50 W, a wettable powder, lists ornamentals *such as* arborvitae, azalea, birch, and so forth. The words *such as* allow the wettable powder formulation to be applied to many ornamentals not named on the label. Be certain that the pesticide product you purchase has a label that allows you to legally treat the plants or other sites you have in mind.

Toxicity of pesticides: common names and trade names. A common indicator of the toxicity of a pesticide to mammals is its LD_{50} value. This value is given in milligrams per kilogram of body weight, and is the amount of the pesticide needed to kill 50 percent of a group of laboratory animals tested, usually rats or rabbits. Both oral and dermal LD_{50} values are shown in table 6 to give an indication of toxicity by accidental ingestion by mouth, and exposure through the skin, respectively. Keep in mind that the **lower** the LD_{50} value, the **more toxic** the pesticide is to mammals.

Oils as insecticides. Most oils available today as insecticides are termed superior-type horticultural oils, and are considered appropriate for use in the dormant season and in the growing (or verdant) season. To date, however, little experimental work has been done by the University of California using oils during the growing season against pests of ornamentals. Detailed discussion on oils as insecticides is available from W. T. Johnson's "Horticultural Oils," *Journal of Environmental Horticulture.*

Oils are contact insecticides that kill certain insects and mites by intervening physically, rather than chemically, with respiratory processes. Research has shown that oils have essentially no residual life; that is, oils affect pests present at the time of application but do not kill pests arriving after the application. Yet, there is some evidence that certain insects arriving soon after treatment may be repelled by the oil residue.

Depending on the pest involved, oils may kill the egg, larval, or adult stage. In the dormant season oils have been useful against scale insects, mites, plant bugs, psyllids, and certain moths.

Whether oils are used in the dormant or growing season, they should never be applied to trees or shrubs stressed by a soil moisture deficit. If used in the growing season, do not make application to plants pushing out the spring flush of growth, for tender foliage may be injured.

Plants reportedly sensitive to oils include maple, hickory, walnut, cryptomeria, smokebush, and azalea. Those having a tendency toward sensitivity to oils include beech, Japanese holly, redbud, photinia, spruce, and Douglas fir. Oils will temporarily remove the glaucus bloom from such conifers as Colorado blue spruce.

Soaps as insecticides. Several insecticidal soap products (such as Safer Agro-Chem's Insecticidal Soap and Acco Highway Spray) are available in California for landscape pest control during the growing (verdant) season. They have some of the same characteristics as horticultural oils in that they kill principally by physical rather than chemical action, and have essentially no residual life, meaning the application will have to be repeated if the target pest has a long period of activity during the growing season. Soaps have given good control of certain exposed pests such as aphids, greenhouse thrips, spider mites, psyllids, and whiteflies.

Table 6. Toxicity of pesticides: common names and trade names

		Oral	**Dermal**
Common name	**Trade names**	**LD_{50} (mg/kg)**	
acephate	Orthene	866-945	10,250
Bacillus thuringiensis	Dipel, Thuricide, Bactospeine, SOK-Bt	—*	—
carbaryl	Sevin	500-800	4,000+
diazinon	Diazinon	300-400	600-2,000+
dicofol	Dicofol	684-809	1,000-1,230
dicrotophos	Bidrin	19-30	112-400
dimethoate	Cygon, De-Fend	215	1,000
endosulfan	Thiodan	30-110	359
ethylene dichloride	EDC	670-890	1,000†
fenbutatin-oxide	Vendex	2,361	2,000
insecticidal soap	Safer Insecticidal Soap, Acco Highway Spray	—	—
lindane	Lindane	88-125	1,000
malathion	Cythion	1,000-1,375	4,100
oil	Volck oil	—	—
resmethrin	SPB-1382	4,240	2,500
trichlorfon	Dylox	150-400	500
chlorpyrifos	Dursban	97-276	2,000

*All dashes mean extremely low in mammalian toxicity.
†Acute vapor toxicity.

The Plant Pest Management Table

The pests named in table 7 are considered to be the most important insects and mites likely to be encountered in California; yet, in some cases relatively minor or obscure insects and mites are noted also. By no means could all arthropods of each plant be included, for there are more than 500 species of insects and mites associated with oaks alone. For information on pests not included, and with the absolute certainty that new pests will invade California from outside the state, use the references on page 82.

For color photographs of pests of landscape ornamentals, refer to the book, *Insects that Feed on Trees and Shrubs*, where many of the insects and mites, or their damage, covered in this manual are illustrated. All the pests named in table 7 are not going to be encountered throughout California. Many pests occur sporadically, or only in one part of the state.

Insecticide or miticide management guidelines given in table 7 tend to be conservative in that some of the newer products in the marketplace are not listed. In some cases these products are no more effective than older compounds. In other instances, local information is not available as to the efficacy or other attributes of these newer products. Importantly, too, some are not named because the U.S. Food and Drug Administration has not established tolerances for the residues left on home garden vegetables and fruit crops when nearby landscape trees are sprayed with insecticides and an inevitable drift occurs.

There is no single "best" recommendation for any pest problem. Each situation must be explored in the light of peculiarities and constraints that exist at the site, and then a decision made carefully. Sometimes the decision will be to take no action.

Occasionally, pesticide use for particular crops, including landscape ornamentals, is cancelled by authority of federal or state agencies, or both. You are urged to double check with your local county Agricultural Commissioner for each use of any pesticide named in this manual, to be certain that its intended application is still lawful.

Table 7. Plant pest management

Symptoms/signs	Pest and description	Management options	Remarks
		Acacia	
Chlorosis or dieback of plant tips. White pellets of honeydew collect on foliage, causing blackening of plant by sooty mold.	ACACIA PSYLLID, *Psylla uncatoides*. Adults are less than ⅛ inch long, tan to brown leafhopper-like insects with membranous wings. Nymphs are smaller, orange to green, flattened insects, and found in leaf axils or clustered on growing points. Suck sap.	Avoid planting susceptible acacia species (see p. 14). In most areas psyllid is under biological control by natural enemies. If not, apply carbaryl or diazinon.	Chemical control is difficult and must be repeated, for insect has about 8 generations each year. Populations are highest in spring. Fewer than 25 insects per tip are not likely to cause noticeable damage.
Stippling (flecking) or bleaching of foliage.	LEAFHOPPER, *Kunzeana kunzii*. Adults are just over 1/16 inch long and dull green. Adults and nymphs suck sap.	Apply carbaryl.	A pest late in growing season. Most damage has been seen on *Acacia decurrens*.
Wet, white, frothy masses of spittle on foliage or twigs.	SPITTLEBUG, *Clastoptera arizonica*. Greenish bugs found within spittle masses. Suck sap.	No noticeable plant injury has been documented.	Pest more common in southern California than northern California.
Blackening of foliage from sooty mold. White, immobile, popcornlike bodies on woody parts. Possibly decline of plant.	COTTONY CUSHION SCALE, *Icerya purchasi*. Adult females produce cottony white egg sacs. Immature scales are orange to brown, flattened, immobile insects on foliage or woody parts. Suck sap.	Normally under biological control by natural enemies. If not, apply carbaryl or diazinon when crawlers are numerous.	Insect breeds year-round.

(continued)

Table 7. Acacia *continued*

Symptoms/signs	Pest and description	Management options	Remarks
Decline or dieback of twigs or branches.	ARMORED SCALES: GREEDY SCALE, *Hemiberlesia rapax;* OLEANDER SCALE, *Aspidiotus nerii;* SAN JOSE SCALE, *Quadraspidiotus perniciosus;* CALIFORNIA RED SCALE, *Aonidiella aurantii.* Immobile, grayish to brownish encrustations on twigs and branches. Individual scales are less than 1/16 inch long and circular to oval in shape. Suck sap, may inject toxic saliva into plant.	Various natural enemies normally prevent these scales from reaching damaging levels. If not, apply oil during dormant season. Or, apply diazinon when crawlers are numerous in spring and summer.	A heavy scale infestation is not likely to be reduced to low levels by a single oil application. Several consecutive years of annual oil sprays probably will be needed. These scales have several generations each year.
Dieback of occasional twig.	LEAD CABLE BORER, *Scobicia declivis.* Adults are ¼ inch long, cylindrical, brown to black beetles. Tunnel into twigs or branches, often starting at crotch.	Prune off affected plant parts. Eliminate beetle breeding sources close to acacia.	Beetles breed in dead oak, maple, eucalyptus, and other hardwoods.
Leaves chewed.	OMNIVOROUS LOOPER, *Sabulodes caberata.* Larvae are yellow, green, or pink with lengthwise stripes of yellow, green, or black. Crawl in "looping" manner. About 1½ inches long when mature.	Apply carbaryl, acephate, or *Bacillus thuringiensis.*	Sporadic pest that only occasionally reaches damaging numbers.
Leaves webbed together with silk. Leaves chewed.	ORANGE TORTRIX, *Argyrotaenia citrana.* Larvae are dirty white with brown head and brown "shield" on back just behind head. Nearly ¾ inch long when mature.	Apply acephate, carbaryl, or *Bacillus thuringiensis.*	

Albizia (Silk Tree; Mimosa)

Symptoms/signs	Pest and description	Management options	Remarks
Chlorosis or dieback of plant tips. White pellets of honeydew collect on foliage, causing blackening of plant by sooty mold.	ACACIA PSYLLID, *Psylla uncatoides.* Adults are less than ⅛ inch long, tan to brown leafhopper-like insects with membranous wings. Nymphs are smaller, orange to green, flattened insects, and found in leaf axils or clustered on growing points. Suck sap.	In most areas psyllid is under biological control by natural enemies. If not, apply carbaryl or diazinon.	Chemical control is difficult and must be repeated, for insect has about 8 generations each year. Populations are highest in spring. Fewer than 25 insects per tip are not likely to cause noticeable damage.

(continued)

Table 7. Albizia *continued*

Symptoms/signs	Pest and description	Management options	Remarks
Tents of silk in tree. Leaves chewed.	MIMOSA WEBWORM, *Homadaula anisocentra.* Larvae are pale gray to dark brown with longitudinal white stripes. About ½ inch long when mature.	Prune off affected branch tips with webs, when first seen and tents are small. Or, apply *Bacillus thuringiensis,* carbaryl, or acephate.	Most common in Sacramento Valley.

Alder

Symptoms/signs	Pest and description	Management options	Remarks
Dieback of branches. Wet spots on bark of branches or trunk. Gnarled, ridged growth on woody parts. "D"-shaped emergence holes in bark, about ⅛ inch in diameter.	FLATHEADED BORER, *Agrilus burkei.* Slender, white, heavily segmented larva about ½ inch long when mature. Found in winding gallery just beneath bark. Larva girdles tree parts. Ridged growth is result of callus growth produced by tree in attempt to repair injury.	Keep tree vigorous by proper irrigation and fertilization practices. Insecticidal controls are under investigation.	Most injury has been seen on poor growing sites; following drought years; and where irrigation is inadequate. Insect has one generation each year, with adults emerging in spring. Italian alder, *Alnus cordata,* is resistant.
Stickiness and blackening of foliage from honeydew and sooty mold. Cast skins on underside of leaves.	APHIDS: *Eucaraphis gillettei; Pterocallis alni.* Yellowish green insects, less than ⅛ inch long, on leaves. Suck sap.	Often controlled by ladybird beetles. If not, apply diazinon, acephate, or insecticidal soap.	
Cottony white tufts of waxy material on leaves.	COTTONY ALDER PSYLLID, *Psylla alni.* Yellow to green insects, about 1/16 inch long, beneath waxy tufts. Suck sap.	Apply diazinon.	
Leaves skeletonized.	ALDER FLEA BEETLE, *Altica ambiens.* Adults are metallic blue and ¼ inch long. Larvae are brown to black.	Apply carbaryl.	Only one generation each year.
Stippling (flecking) or bleaching of leaves. Specks of varnishlike material and cast skins on underside of leaves.	LACE BUGS, *Corythucha* spp. Adults are about ⅛ inch long, flattened, with lacy wings. Nymphs are smaller and spiny. Suck sap from underside of leaves.	Apply carbaryl.	

(continued)

Table 7. Alder *continued*

Symptoms/signs	Pest and description	Management options	Remarks
Stickiness and blackening of foliage from honeydew and sooty mold. Possibly some decline or dieback of woody parts.	SOFT (UNARMORED) SCALES: COTTONY MAPLE SCALE, *Pulvinaria innumerabilis;* EUROPEAN FRUIT LECANIUM, *Parthenolecanium corni.* Brown, flattened or hemispherical, immobile bodies, less than ¼ inch long, on twigs. Cottony maple scale females, when mature, look like popcorn. Suck sap.	Normally held in check by natural enemies. If not, apply carbaryl or diazinon when crawlers are numerous in late spring to early summer.	Cottony maple scale has been most noticeable in foothill and mountainous areas of northern California. Both scale insects have one generation each year.
Stunting or dieback of woody parts.	OYSTERSHELL SCALE, *Lepidosaphes ulmi.* Gray to brown immobile encrustations on twigs and branches. Individual scales about 1/16 inch long, look like miniature oysters. Suck sap, may inject toxic saliva into plant.	Various natural enemies normally prevent this scale from reaching damaging levels. If not, apply oil during dormant season. Or, apply diazinon when crawlers are numerous in spring and summer.	A heavy scale infestation is not likely to be reduced to low levels by a single oil application. Several consecutive years of annual oil sprays probably will be needed. Insect has 1 generation each year in northern California; 2 generations in southern California.

Arborvitae

Symptoms/signs	Pest and description	Management options	Remarks
Browning of tips of growth. Browning begins in fall and is at its worst in late winter to spring.	CYPRESS TIP MINER, *Argyresthia cupressella.* Adults are silvery tan moths with wingspan about ¼ inch. Larvae are green, about ¼ inch long when mature, and tunnel inside foliage.	Plant resistant arborvitae (see p. 14). Apply acephate, diazinon, or carbaryl when moths are active, usually March to April in southern California, and April to May in northern California.	To determine when moths are active, shake foliage beginning in early March. Moths will fly out, then return to foliage. One annual generation. Insect occurs along coast only.
Stickiness and blackening of foliage from honeydew and sooty mold.	ARBORVITAE APHID, *Dilachnus tujafilinus.* Brown to gray, ⅛ inch long insects on foliage and twigs. Suck sap.	Apply acephate, diazinon, or insecticidal soap.	
Discoloration and possible dieback of growth. Circular to oval immobile bodies, about 1/16 inch long, on foliage or shoots.	ARMORED SCALES: JUNIPER SCALE, *Carulaspis juniperi;* MINUTE CYPRESS SCALE, *C. minima.* Individual scales are circular to oval, immobile, and about 1/16 inch long. Suck sap.	Populations in California have very seldom warranted control. If control is necessary, apply diazinon when crawlers are numerous in spring to summer.	Insects have several generations each year.

(continued)

Table 7. Arborvitae *continued*

Symptoms/signs	Pest and description	Management options	Remarks
Lateral twigs killed for distance of about 6 inches back from their tips. Dead tips, or "flags," hang on tree.	CEDAR OR CYPRESS BARK BEETLES, *Phloeosinus* spp. Adults are dark brown to black beetles about ⅛ inch long. Larvae are white legless grubs found in tunnels beneath bark. Adults mine tips of twigs, causing "flags." Also bore through bark of trunks of declining arborvitae. Eggs laid in galleries there. Plant is girdled.	Ignore, or prune off, flags. Maintain plants in good vigor to reduce likelihood of trunk and limb attack.	Flagging of tips does not mean adults are attacking trunk or limbs of the same plant.
Branches killed back, sometimes to trunk. Coarse boring dust found in branch crotches, or at point where trunk may have been physically wounded.	CYPRESS BARK MOTH, *Laspeyresia cupressana.* Larvae are pink to gray and about ½ inch long when mature. Found beneath bark.	Control of insect is seldom warranted. Dieback of limbs or entire plant may be due to cypress canker, a disease. Bark moth larvae are secondary, often attacking at sites of cankers or physical wounds.	*Platycladus* (=*Thuja*) *orientalis* is reported to be a host of cypress canker.

Ash

Symptoms/signs	Pest and description	Management options	Remarks
Coarse, yellow stippling (flecking) of leaves. Trees may be defoliated.	ASH PLANT BUGS, *Tropidosteptes illitus* and *T. pacificus.* Yellow-brown bugs about 3/16 inch long when mature. Suck sap.	Apply carbaryl.	By June or July bugs have completed their development. Eggs are laid in twigs and will hatch the next spring. Most destructive in Central Valley.
Stippling (flecking) of leaves. Dark specks of excrement on underside of leaves.	LACE BUG, *Leptoypha minor.* Pale brown insects, less than ⅛ inch long at maturity, with lacelike wings. Suck sap.	Apply carbaryl.	4 to 5 annual generations. Damage is most prominent in late summer.
Curled, twisted, "galled" leaves on branch tips. Sticky honeydew on surfaces below tree.	APHID, *Prociphilus californicus.* Pale green or grayish insects on terminal leaves or inside "galled" foliage. Suck sap.	Apply acephate or diazinon.	For maximum effect, sprays should be applied in spring before foliage becomes "galled."
Branches wilting or dying. Boring dust in crotches or other woody parts. Swollen areas and holes on trunk or branches.	ASH/LILAC BORER, *Podosesia syringae.* Larva is creamy white with brown head and about 1 inch long when mature. Found beneath bark or in heartwood.	Management under investigation.	At present insect is found in Sacramento, San Joaquin, and Stanislaus counties. Reported from Moraine ash, Raywood ash, *Fraxinus oxycarpa*, and Modesto ash, *F. velutina.* Not reported from evergreen or shamel ash, *F. uhdei.* Also reported from lilac and olive. Insect has 1 annual generation, with adults emerging in spring.

(continued)

Table 7. Ash *continued*

Symptoms/signs	Pest and description	Management options	Remarks
Decline, or dieback of twigs and branches.	WALNUT SCALE, *Quadraspidiotus juglansregiae.* Tan or gray encrustations on twigs and branches. Individual scales are 3/16 inch in diameter. Suck sap.	Apply oil-diazinon combination late in dormant season. Or, apply diazinon when crawlers are emerging in spring and summer.	A problem mostly in central and southern San Joaquin Valley. Insect has several generations each year.

Azalea

Symptoms/signs	Pest and description	Management options	Remarks
Wilting and death of plants. Roots may be missing or their bark removed, or plant may be girdled at or just below soil surface.	ROOT WEEVILS: BLACK VINE WEEVIL, *Otiorhynchus sulcatus;* WOODS WEEVIL, *Nemocestes incomptus.* Adults are dark brown to black snout beetles, ¼ to ½ inch long. Larvae occur in soil and are white legless grubs with brown head.	Apply acephate to plants and soil surface beginning when "scalloping" or notching of leaf margins occurs on nearby rhododendron, photinia, pyracantha, and others. Sprays are intended to kill adult weevils before they lay the eggs that become the more destructive larval stage.	Depending on root weevil species, "scalloping" of margins of leaves may begin from April to July. If adult control is not achieved, be advised that control of larvae in soil is extremely difficult. If possible, spray before or after bloom, for sprays may injure open blossoms.
Browning of leaves, or leaves tied together.	AZALEA LEAFMINER, *Gracillaria azaleella.* Larvae are green, ½ inch long when mature, and difficult to find because they are secretive.	Apply acephate or carbaryl.	If possible, spray before or after bloom, for sprays may injure open blossoms. Larvae are leafminers when young, and feed externally when maturing.

Baccharis

Symptoms/signs	Pest and description	Management options	Remarks
Swellings (galls) on tips of shoots.	GALL FLY, *Rhopalomyia californica.* Orange maggots, about ⅛ inch long when mature, inside galls. Galls stop shoot growth.	Management never investigated.	Insect has many generations each year.
Blackening of foliage or woody parts as a result of sooty mold. Possibly decline of plant.	BLACK SCALE, *Saissetia oleae.* Adults are immobile, dark brown to black bulbous insects, ⅛ to 3/16 inch long, with raised letter "H" on back. Immature scales are orange to brown, oval, flattened insects. Occur on woody parts and sometimes leaves. Suck sap.	Often under adequate biological control by natural enemies. If not, apply carbaryl or diazinon when crawlers are numerous.	Insect has 2 generations each year along the coast and 1 annual generation inland.

(continued)

Table 7. Baccharis *continued*

Symptoms/signs	Pest and description	Management options	Remarks
Decline of plant.	GREEDY SCALE, *Hemiberlesia rapax*. Immobile, grayish encrustations on woody parts. Individual scales are less than 1/16 inch long and nearly circular in shape. Suck sap, may inject toxic saliva into plant.	Various natural enemies normally prevent this scale from reaching damaging levels. If not, apply oil during dormant season. Or, apply diazinon when crawlers are numerous in spring and summer.	A heavy scale infestation is not likely to be reduced to low levels by a single oil application. Several consecutive years of annual oil sprays probably will be needed. Insect has several generations each year.
Chewed foliage. Plant may be defoliated.	LOOPER, *Prochoerodes truxaliata*. Brown to purplish caterpillars, 1½ inches long when mature.	Apply acephate, carbaryl, or *Bacillus thuringiensis* when larvae are small.	Several generations each year. Larvae are difficult to find; they descend to base of plant in daytime, crawl up to feed at night. Omnivorous looper may also defoliate plants (see IVY).
Chewed foliage. Plant may be defoliated.	BACCHARIS LEAF BEETLE, *Trirhabda flavolimbata*. Adult beetle has yellowish head, metallic blue or green back. Larvae are blackish brown and ½ inch long when mature.	Apply carbaryl.	Insect has one generation each year. Damage is done in spring. Beetles tend to drop off plant when approached.
Stippling (flecking) of leaves. Foliage bleached. Varnishlike specks of excrement and cast skins on underside of leaves.	LACE BUG, *Corythucha* (probably) *morrilli*. Adults are brownish with lacelike wings and about ⅛ inch long. Suck sap.	Apply carbaryl.	
Decline or dieback of branches or entire plant. Tunneling in woody parts.	FLATHEADED BORER, *Chrysobothris* (probably) *bacchari*. Horseshoe nail-shaped white larvae, about 1¼ inches long when mature, tunnel in woody party.	Prune off and destroy affected parts. Insecticidal controls not tested.	

Bamboo

Symptoms/signs	Pest and description	Management options	Remarks
Yellowing of leaves. Sticky and blackened plants from honeydew and sooty mold. Cast skins present.	BAMBOO APHIDS, *Takecallis* spp. Pale yellow insects with black markings, about 1/16 inch long, found on leaves. Suck sap.	Apply diazinon, acephate, or insecticidal soap.	
Dieback of foliage. Cottony material in leaf axils. Blackening of foliage from sooty mold.	MEALYBUGS (several species). Purplish, heavily segmented insects, up to 3/16 inch long, found beneath mealy gray or white material. Suck sap.	Apply diazinon or malathion.	Addition of extra wetting agent will improve penetration of mealy material covering mealybugs.

(continued)

Table 7. Bamboo *continued*

Symptoms/signs	Pest and description	Management options	Remarks
Dieback of plant parts.	BAMBOO SCALE, *Asterolecanium bambusae.* Yellowish to blackish, oval, immobile insects about 1/16 inch long. Found on leaves and stems, often under sheaths. Suck sap.	Apply oil in dormant period. Or, apply diazinon or carbaryl several times, 3 weeks apart, in spring when crawlers appear.	Pest of minor importance in landscape.

Bay (California Bay Laurel)

Symptoms/signs	Pest and description	Management options	Remarks
Sticky, blackened foliage from honeydew and sooty mold.	CALIFORNIA LAUREL APHID, *Euthoracaphis umbellulariae.* Powdery gray immobile bodies, about 1/16 inch long, on undersides of leaves, often along major veins. Suck sap.	Although blackening is unsightly, no harm to tree results from even a heavy infestation. Apply insecticidal soap sprays to reduce aphid and sooty mold levels.	This aphid is often mistaken for a whitefly or scale insect.
Twig and branch dieback.	ARMORED SCALES, including GREEDY SCALE, *Hemiberlesia rapax;* OLEANDER SCALE, *Aspidiotus nerii.* Immobile, grayish to brownish encrustations on twigs and branches. Individual scales are less than 1/16 inch long and circular to oval in shape. Suck sap, may inject toxic saliva into plant.	Various natural enemies normally prevent these scales from reaching damaging levels. If not, apply oil during dormant season. Or, apply diazinon when crawlers are numerous in spring and summer.	A heavy scale infestation is not likely to be reduced to low levels by a single oil application. Several consecutive years of annual oil sprays probably will be needed. These scales have several generations each year.
Foliage blackened from sooty mold.	BLACK SCALE, *Saissetia oleae.* Adults are immobile, dark brown to black, bulbous insects, 1/8 to 3/16 inch long, with raised letter "H" on back. Immature scales are orange to brown, oval, flattened insects. Occur on woody parts and sometimes leaves. Suck sap.	Often under adequate biological control by natural enemies. If not, apply carbaryl or diazinon when crawlers are numerous. Thinning growth, permitting entry of more light and better air circulation, creates environment less favorable for scale.	Insect has 2 generations each year along the coast and 1 annual generation inland.
Dieback of occasional twigs.	BRANCH AND TWIG BORERS, *Scobicia declivis* and *Polycaon confertus.* Adults are 1/4 to 1/2 inch long, brown to black beetles. Tunnel into twigs.	Prune off affected plant parts. Eliminate beetle breeding sources close to bay trees.	Beetles breed in dead oak, maple, bay, eucalyptus, and other hardwoods.

Table 7. *Continued*

Symptoms/signs	Pest and description	Management options	Remarks
	Birch		
Sticky, blackened foliage from honeydew and sooty mold. Cast skins on underside of leaves. Some yellowing of leaves may be evident.	APHIDS, including *Euceraphis betulae; Calaphis betulaecolens.* Yellowish to greenish insects, about 1/16 inch long, on undersides of leaves. Suck sap.	Apply acephate, diazinon, or insecticidal soap.	
Stippling (flecking) of foliage. Leaves may appear bleached. Cast skins on underside of leaves.	LEAFHOPPERS, including *Empoasca* sp.; *Alebra albostriella.* Green, wedge-shaped insects, about ⅛ inch long, on undersides of leaves. Move quickly when disturbed. Suck sap.	Apply carbaryl.	
Blackening of foliage from sooty mold.	FROSTED SCALE, *Parthenolecanium pruinosum.* Convex, brown, hemispherical, immobile bodies on twigs. About ¼ inch long when mature, and covered with whitish, frosting-like wax. Suck sap.	Usually held in check by natural enemies. If not, apply carbaryl for crawler control in summer to early winter.	Insect has 1 generation each year.
Dieback of branches, top of tree, or entire tree.	BORER, *Paranthrene robiniae* Larva is dirty white and 1¼ inches long when mature. Tunnels in sapwood and heartwood of branches or trunk.	As protectant, apply lindane to woody parts of tree in spring. To kill established borers, inject ethylene dichloride into gallery entrance holes, then plug hole entrance.	Adult insect is a clearwing moth that resembles a yellow-jacket wasp.
	Boxelder		
Spotting or yellowing of foliage. More severe on female, than male, trees.	WESTERN BOXELDER BUG, *Leptocoris rubrolineatus.* Adults are ½ inch long and gray with prominent red diagonal markings. Suck sap.	Eliminate female boxelder trees.	Insect is more important as an invading household pest than as a pest of landscape trees.
Blackening of foliage from sooty mold.	COTTONY MAPLE SCALE, *Pulvinaria innumerabilis.* Immature scales are immobile, yellowish or tan, flattened bodies, up to about 1/16 inch long, on leaves. Mature females occur on twigs and look like popcorn. Suck sap.	Apply carbaryl in early summer when crawlers are numerous.	Most often seen in foothill areas and mountainous northern California. Insect has 1 generation each year.

(continued)

Table 7. Boxelder *continued*

Symptoms/signs	Pest and description	Management options	Remarks
Dieback of twigs or limbs.	OYSTERSHELL SCALE, *Lepidosaphes ulmi.* Gray to brown immobile encrustations on woody parts. Individual scales, about 1/16 inch long, look like miniature oysters. Suck sap, may inject toxic saliva into plant.	Various natural enemies normally prevent scale from reaching damaging levels. If not, apply oil during dormant season. Or, apply diazinon when crawlers are numerous in spring and early summer.	A heavy scale infestation is not likely to be reduced to low levels by a single oil application. Several consecutive years of annual oil sprays probably will be needed. Insect has 1 generation each year in northern California; 2 in southern California.
Dieback of woody parts.	FLATHEADED APPLE TREE BORER, *Chrysobothris femorata.* Larvae are dirty white and shaped like horseshoe nails. About 1 inch long when mature. Mine beneath bark or into heartwood.	Take steps to return tree to a state of good vigor.	Insect attacks weakened trees or trees in a declining state of health.
Chewed foliage. Leaves rolled and tied together with silk.	FRUITTREE LEAFROLLER, *Archips argyrospilus.* Larvae are green with black head and shield on back behind head. About ¾ inch long when mature. Wriggle vigorously when touched.	Apply carbaryl, diazinon, or *Bacillus thuringiensis.*	One generation each year. An early season problem.

Boxwood

Symptoms/signs	Pest and description	Management options	Remarks
Cupping of leaves.	BOXWOOD PSYLLID, *Psylla buxi.* Adults are ⅛ inch long, greenish, and have membranous wings. Nymphs are greenish, flattened insects that produce waxy secretions. Suck sap.	Not a serious problem.	American boxwood is more susceptible than English boxwood.
Stunted growth or dieback of twigs and branches.	ARMORED SCALES: GREEDY SCALE, *Hemiberlesia rapax;* OLEANDER SCALE, *Aspidiotus nerii;* OYSTERSHELL SCALE, *Lepidosaphes ulmu;* CALIFORNIA RED SCALE, *Aonidiella aurantii.* Immobile, grayish to brownish encrustations on twigs and branches. Individual scales are less than 1/16 inch long and circular, oval, or elongate in shape. Suck sap, may inject toxic saliva into plant.	Various natural enemies normally prevent these scales from reaching damaging levels. If not, apply oil during dormant season. Or, apply diazinon when crawlers are numerous in spring and summer.	A heavy scale infestation is not likely to be reduced to low levels by a single oil application. Several consecutive years of annual oil sprays probably will be needed. These scales have several generations each year.

(continued)

Table 7. Boxwood *continued*

Symptoms/signs	Pest and description	Management options	Remarks
Blackening of foliage from sooty mold. White, immobile, popcornlike bodies on woody parts. Possibly decline of plant.	COTTONY CUSHION SCALE, *Icerya purchasi*. Adult females produce cottony white egg sacs. Immature scales are orange to brown, flattened, immobile insects on foliage or woody parts. Suck sap.	Normally under biological control by natural enemies. If not, apply carbaryl or diazinon when crawlers are numerous.	Insect breeds year-round.
Broom			
Stickiness and blackening of foliage from honeydew and sooty mold. Twisting of terminal growth.	BEAN APHID, *Aphis fabae*. Dull black insects, less than ⅛ inch long, clustered on growing points or leaves. Suck sap.	Apply acephate, diazinon, or insecticidal soap.	
Chewed leaves. Plant may be defoliated.	GENISTA CATERPILLAR, *Uresiphita reversalis*. Caterpillars are orange-green with black and white hairs, and reach 1 to 1¼ inches when mature.	Apply *Bacillus thuringiensis*, carbaryl, or acephate.	More common in southern California than northern California.
Dieback of plant parts.	ARMORED SCALES: GREEDY SCALE, *Hemiberlesia rapax;* OLEANDER SCALE, *Aspidiotus nerii;* OYSTERSHELL SCALE, *Lepidosaphes ulmi*. Immobile, grayish to brownish encrustations on twigs. Individual scales are less than 1/16 inch long and circular, oval, or elongate in shape. Suck sap, may inject toxic saliva into plant.	Various natural enemies normally prevent these scales from reaching damaging numbers. If not, apply oil during dormant season. Or, apply diazinon when crawlers are numerous in spring and summer.	A heavy scale infestation is not likely to be reduced to low levels by a single oil application. Several consecutive years of annual oil sprays probably will be needed. These scales have several generations each year.

Table 7. *Continued*

Symptoms/signs	Pest and description	Management options	Remarks
		Cactus	
Dieback of plant parts. Whitish to brownish encrustations on plant.	ARMORED SCALES: CACTUS SCALE, *Diaspis echinocacti;* GREEDY SCALE, *Hemiberlesia rapax;* OLEANDER SCALE, *Aspidiotus nerii.* Immobile whitish, grayish, or brownish encrustations on plant. Individual scales are less than 1/16 inch long and circular to oval in shape. Suck sap, may inject toxic saliva into plant.	Apply malathion when crawlers are numerous in spring and summer.	These scales have several generations each year.
Blackening of plant from sooty mold. Cottony waxy material on plant.	LONGTAILED MEALYBUG, *Pseudococcus longispinus.* Powdery white insects about 1/8 inch long when mature. Filaments at tail end are as long or longer than body. Suck sap.	Normally under biological control by natural enemies. If not, apply diazinon or malathion.	
		Camellia	
Blackening of foliage from sooty mold. Leaf cupping, curling, or twisting.	BLACK CITRUS APHID, *Toxoptera aurantii.* Brown to black insects, 1/16 inch long, clustered on buds, leaves, or growing points. Suck sap.	Normally controlled by natural enemies but sometimes not before damage is done. Apply diazinon, acephate, or insecticidal soap.	
Stunting, poor growth, or dieback of twigs or branches.	ARMORED SCALES, including: GREEDY SCALE, *Hemiberlesia rapax;* OLEANDER SCALE, *Aspidiotus nerii;* OYSTERSHELL SCALE, *Lepidosaphes ulmi.* Immobile, grayish to brownish encrustations on twigs or branches. Individual scales are less than 1/16 inch long and circular, oval, or elongate in shape. Suck sap, may inject toxic saliva into plant.	Various natural enemies normally prevent these scales from reaching damaging levels. If not, apply oil during dormant season. Or, apply diazinon when crawlers are numerous in spring and summer.	A heavy scale infestation is not likely to be reduced to low levels by a single oil application. Several consecutive years of annual oil sprays probably will be needed. These scales have several generations each year.

(continued)

Table 7. Camellia *continued*

Symptoms/signs	Pest and description	Management options	Remarks
Blackening of foliage from sooty mold. Possibly decline or dieback of twigs or branches.	SOFT (UNARMORED) SCALES: BROWN SOFT SCALE, *Coccus hesperidum;* BLACK SCALE, *Saissetia oleae.* Immature scales are orange, yellow, or brown, oval, flattened, immobile insects up to 1/16 inch long. Adult brown soft scale remains flattened, brown, immobile insect slightly larger than 1/8 inch. Adult black scale is brown to black, bulbous, with raised letter "H" on back. Suck sap.	Normally controlled by natural enemies. If not, apply carbaryl or diazinon whenever crawlers are numerous.	Brown soft scale has 3 to 5 overlapping generations each year, hence all stages are present most of year. Black scale has 2 generations each year along the coast and 1 annual generation inland.
Chewed leaves and blossoms.	FULLER ROSE BEETLE, *Pantomorus cervinus.* Adults are pale brown weevils with blunt snout and about 3/8 inch long.	Apply acephate.	

Ceanothus

Symptoms/signs	Pest and description	Management options	Remarks
Spindle-shaped swellings (galls) on green stems. Reduced flowering of plant.	CEANOTHUS STEM GALL MOTH, *Periploca ceanothiella.* Larva is gray, about 1/4 inch long, and found inside gall.	Insect is most common on *Ceanothus griseus* and its cultivars such as 'Horizontalis.' Avoid planting these cultivars. Clipping and destroying galled twigs before adults emerge in spring may be helpful.	Insecticidal controls not tested.
Stippling (flecking) of leaves. Foliage may become bleached. Dark specks of excrement on underside of leaves.	CEANOTHUS TINGID (LACE BUG), *Corythucha obliqua.* Adults are brown, nearly 3/16 inch long, with lacelike wings. Nymphs are smaller, flattened insects. Suck sap from underside of leaves.	Apply carbaryl.	
Cottony spots up to 1/4 inch long on underside of leaves.	PSYLLID, *Euphalerus vermiculosus.* Greenish nymphs, 1/16 to 1/8 inch long, found beneath cottony material. Suck sap.	Management never investigated.	No apparent injury is caused by this insect.
Blackening of foliage from sooty mold. Reduced shoot growth.	CEANOTHUS APHID, *Aphis ceanothi.* Reddish brown to black insects, 1/16 inch long, clustered on leaves and shoots. Suck sap.	Usually under biological control by natural enemies. If not, apply diazinon, acephate, or insecticidal soap.	

(continued)

Table 7. Ceanothus *continued*

Symptoms/signs	Pest and description	Management options	Remarks
Dieback of twigs or branches.	ARMORED SCALES, including: OYSTERSHELL SCALE, *Lepidosaphes ulmi;* GREEDY SCALE, *Hemiberlesia rapax.* Immobile, grayish to brownish encrustations on twigs or branches. Individual scales are less than 1/16 inch long and circular to elongate in shape. Suck sap, may inject toxic saliva into plant.	Various natural enemies normally prevent these scales from reaching damaging levels. If not, apply oil during dormant season. Or, apply diazinon when crawlers are numerous in spring and summer.	A heavy scale infestation is not likely to be reduced to low levels by a single oil application. Several consecutive years of annual oil sprays probably will be needed.

Cedar (Cedrus)

Symptoms/signs	Pest and description	Management options	Remarks
Stickiness and blackening of foliage from honeydew and sooty mold.	GIANT CONIFER APHID, *Cinara* spp. Dark, long-legged insects about 1/8 inch long on limbs. Suck sap.	Apply acephate, diazinon, or insecticidal soap.	More common in southern, than northern, California. Sometimes only a single limb is infested.

Chamaecyparis (False Cypress)

Symptoms/signs	Pest and description	Management options	Remarks
Browning of growth tips. Browning begins in fall and is at its worst in late winter to spring.	CYPRESS TIP MINER, *Argyresthia cupressella.* Adults are silvery tan moths with wingspan about 1/4 inch. Larvae are green, about 1/4 inch long when mature, and tunnel inside foliage.	Apply acephate, diazinon, or carbaryl when moths are active, usually March to April in southern California and April to May in northern California.	To determine when moths are active, shake foliage beginning in early March. Moths will fly out, then return to foliage. One annual generation. Insect occurs along coast only.
Stickiness and blackening of foliage from honeydew and sooty mold.	ARBORVITAE APHID, *Dilachnus tujafilinus.* Brown to gray, 1/8 inch long insects on foliage and twigs. Suck sap.	Apply acephate, diazinon, or insecticidal soap.	
Yellowing or browning of foliage. Whitish to brownish encrustations on foliage.	JUNIPER SCALE, *Carulaspis juniperi.* Individual scales are circular to elongate, immobile, and about 1/16 inch long. Suck sap.	Populations in California have very seldom warranted control. If control is necessary, apply diazinon when crawlers are numerous in spring and summer.	Insect has several generations each year.
Bits of dead and living foliage tied together with silk. Plant may show browning.	CYPRESS LEAF TIER, *Epinotia subviridis.* Blackish pink larvae, 3/8 inch long when mature. Feed from protection of tubular "nests."	Apply spray of carbaryl or diazinon at high pressure. March to April is normal time for treatment.	Also occurs on cypress, arborvitae, cedar, and juniper. More of a pest in southern California than northern California.

(continued)

Table 7. Chamaecyparis *continued*

Symptoms/signs	Pest and description	Management options	Remarks
Stippling (flecking) of foliage. Foliage may be yellowed.	SPRUCE SPIDER MITE, *Oligonychus ununguis.* Greenish specks about the size of ground pepper. Fine webbing often present on foliage. Suck sap from foliage.	First determine presence of increasing mite numbers by periodic shaking of foliage over a white surface. Apply fenbutatin-oxide.	A "cool weather" mite. Highest populations occur in spring and fall.
Twigs killed back for distance of about 6 inches from their tips. Dead tips, or "flags," hang on tree.	CEDAR OR CYPRESS BARK BEETLES, *Phloeosinus* spp. Adults are dark brown to black beetles about ⅛ inch long. Larvae are white legless grubs found in tunnels beneath bark. Adults mine tips of twigs, causing "flags." Also bore through bark of trunks of declining Chamaecyparis. Eggs laid in galleries there. Plant is girdled.	Ignore, or prune off, flags. Maintain plants in good vigor to reduce likelihood of trunk and limb attack.	Flagging of tips does not mean adults are attacking trunk or limbs of the same plant.
Branches killed back, sometimes to trunk. Coarse boring dust found in branch crotches, or at point where trunk may have been physically wounded.	CYPRESS BARK MOTH, *Laspeyresia cupressana.* Larvae are pink to gray and about ½ inch long when mature. Found beneath bark.	Control of insect is seldom warranted. Dieback of limbs or entire plant may be due to cypress canker, a disease. Bark moth larvae are secondary, often attacking at sites of cankers or physical wounds.	Chamaecyparis is only an occasional host of cypress canker.

Citrus

Symptoms/signs	Pest and description	Management options	Remarks
Blackening of foliage from sooty mold. Possibly poor growth of tree.	SOFT (UNARMORED) SCALES, including: BROWN SOFT SCALE, *Coccus hesperidum;* BLACK SCALE, *Saissetia oleae.* COTTONY CUSHION SCALE, *Icerya purchasi.* Immature scales are orange to brown, oval, flattened, immobile insects on foliage or twigs. Adult brown soft scale is flattened, brown insect. Adult black scale is brown to black, bulbous, with raised letter "H" on back. Adult cottony cushion scale produces cottony white egg sac. Suck sap.	Normally under adequate to good biological control by natural enemies. If not, apply carbaryl or diazinon when crawlers are active. Oil may be used for scale control during growing season.	Brown soft scale and cottony cushion scale have several generations each year. Black scale has 2 generations each year along the coast and 1 annual generation inland.
Stickiness and blackening of foliage from honeydew and sooty mold. Cast skins on underside of leaves.	APHIDS, including: BLACK CITRUS APHID, *Toxoptera aurantii;* SPIREA APHID, *Aphis citricola.* Green to brownish insects, 1/16 inch long. Cluster on growing points and underside of leaves. Suck sap.	Apply diazinon or insecticidal soap.	More common in coastal areas. Treatments often not needed in interior valleys.

(continued)

Table 7. Citrus *continued*

Symptoms/signs	Pest and description	Management options	Remarks
Blackening of foliage from sooty mold. Cottony waxy material on plant.	MEALYBUGS, *Pseudococcus* and *Planococcus* spp. Powdery gray insects, not more than ¼ inch long when mature. Fringe of waxy filaments around edge of body. Suck sap.	Usually under biological control by natural enemies. If not, apply malathion or diazinon.	
Blackening of foliage from sooty mold.	WHITEFLIES, including: GREENHOUSE WHITEFLY, *Trialeurodes vaporariorum;* WOOLLY WHITEFLY, *Aleurothrixus floccosus;* CITRUS WHITEFLY, *Dialeurodes citri.* Adults are about 1/16 inch long with powdery white wings. Immature whiteflies are immobile, flattened, yellowish to greenish oval bodies, up to 1/16 inch long, on underside of leaves. Suck sap.	Apply malathion or insecticidal soap.	For best control, make several applications 4 to 6 days apart.
Stunting, decline, or dieback of parts of plant.	ARMORED SCALES, including: CALIFORNIA RED SCALE, *Aonidiella aurantii;* YELLOW SCALE, *A. citrina;* PURPLE SCALE, *Lepidosaphes beckii.* Immobile, circular to elongage bodies, not more than 1/16 inch long, on leaves and woody parts. Appear as grayish to orange-brown encrustations. Suck sap, may inject toxic saliva into plant.	Often under good biological control by natural enemies, especially in coastal areas. If not, apply oil during growing season. Or, apply malathion when crawlers are active.	These scales have several generations each year.
Stippling (flecking) of leaves. Foliage may appear bleached. Leaves may drop.	SPIDER MITES, including: CITRUS RED MITE, *Panonychus citri;* PACIFIC SPIDER MITE, *Tetranychus pacificus;* TWOSPOTTED SPIDER MITE, *T. urticae.* Tiny specks about the size of ground pepper, on leaves. Suck sap.	Apply oil.	
Leaves chewed. Leaves tied together with silk.	CATERPILLARS AND LEAFROLLERS, including: ORANGE TORTRIX, *Argyrotaenia citrana;* FRUITTREE LEAFROLLER, *Archips argyrospilus;* WESTERN TUSSOCK MOTH, *Orgyia vetusta.* Naked to hairy larvae. Some are 1 inch long at maturity.	Apply *Bacillus thuringiensis* or carbaryl.	If carbaryl is used, add a miticide to the spray tank to prevent a possible spider mite problem.

Table 7. *Continued*

Cotoneaster

Symptoms/signs	Pest and description	Management options	Remarks
Stippling (flecking) of leaves. Foliage may become bleached. Dark specks of excrement and cast skins on underside of leaves.	TINGID (LACE BUG), probably *Corythucha* sp. Adults are pale brown, about 1/8 inch long, with lacelike wings. Nymphs are smaller, flattened insects. Suck sap.	Apply carbaryl.	
Twig dieback and decline of plant.	ARMORED SCALES: OYSTERSHELL SCALE, *Lepidosaphes ulmi;* GREEDY SCALE, *Hemiberlesia rapax;* SAN JOSE SCALE, *Quadraspidiotus perniciosus.* Immobile, grayish to brownish encrustations on twigs and branches. Individual scales are less than 1/16 inch long and circular, oval, or elongate in shape. Suck sap, may inject toxic saliva into plant.	Various natural enemies normally prevent these scales from reaching damaging levels. If not, apply oil during dormant season. Or, apply diazinon when crawlers are numerous in spring and summer.	These scales have several generations each year.
Stickiness and blackening of plant from honeydew and sooty mold. Cast skins on plant.	APPLE APHID, *Aphis pomi.* Bright green insects, about 1/16 inch long, clustered on growing points or leaves. Suck sap.	Often under biological control by natural enemies. If not, apply acephate, diazinon, or insecticidal soap.	

Cypress

Symptoms/signs	Pest and description	Management options	Remarks
Browning of growth tips. Browning begins in fall and is at its worst in late winter to spring.	CYPRESS TIP MINER, *Argyresthia cupressella.* Adults are silvery tan moths with wingspan about 1/4 inch. Larvae are green, about 1/4 inch long when mature, and tunnel inside foliage.	Apply acephate, diazinon, or carbaryl when moths are active, usually about March to April in southern California and April to May in northern California.	To determine when moths are active, shake foliage beginning in early March. Moths will fly out, then return to foliage. One annual generation. Insect occurs along coast only.
			Monterey cypress is the most susceptible cypress.
Discoloration of foliage. Circular to oval immobile bodies, about 1/16 inch long, on foliage or shoots.	ARMORED SCALES: JUNIPER SCALE, *Carulaspis juniperi;* MINUTE CYPRESS SCALE, *C. minima.* Individual scales are circular to oval, immobile, and about 1/16 inch long. Suck sap.	Populations in California have very seldom warranted control. If control is necessary, apply diazinon when crawlers are numerous in spring and summer.	Insects have several generations each year.

(continued)

Table 7. Cypress *continued*

Symptoms/signs	Pest and description	Management options	Remarks
Lateral twigs killed for distance of about 6 inches back from their tips. Dead tips, or "flags," hang on tree.	*CEDAR OR CYPRESS BARK BEETLES, Phloeosinus* spp. Adults are dark brown to black beetles about ⅛ inch long. Larvae are white legless grubs found in tunnels beneath bark. Adults mine tips of twigs, causing "flags." Also bore through bark of trunks of declining trees. Eggs laid in galleries there. Larvae mine beneath bark. Tree is girdled.	Ignore, or prune off, flags. Maintain trees in good vigor to reduce likelihood of trunk and limb attack.	Flagging of tips does not mean adults are attacking trunk or limbs of the same tree.
Coarse boring dust in branch crotches, on trunk, at point where trunk may have been physically wounded, or on cone clusters. Bleeding of trunk or limbs.	CYPRESS BARK MOTH, *Laspeyresia cupressana.* Larvae are pink to gray and about ½ inch long when mature. Found beneath bark or in cones.	Control of insect is seldom warranted. Dieback of limbs or entire tree is often due to cypress canker, a disease. Bark moth larvae are secondary, often attacking at sites of cankers or physical wounds.	Monterey cypress is the most common host of this insect.
Bits of dead or living foliage tied together with silk to form "nests." Plant may show browning.	CYPRESS WEBBER, *Herculia phoezalis;* and CYPRESS LEAF TIER, *Epinotia subviridis.* Dark-colored larvae, ⅜ to ¾ inch long when mature, feed from protection of "nests."	Apply spray of carbaryl or diazinon at high pressure. March to April is normal time for treatment.	More of a pest in southern California than northern California. Most common on Italian and Monterey cypress.
Tufts of cottony material protruding from cracks and beneath bark flakes on trunks and branches. Possibly some decline in tree.	CYPRESS BARK MEALYBUG, *Ehrhornia cupressi.* Mostly immobile, reddish insects, about 1/16 inch long, found beneath cottony material. Suck sap.	Normally under biological control by natural enemies. If not, apply diazinon spray at high pressure.	Also known as cypress bark scale. A serious problem early in 20th century, less so today.

Douglas Fir

Symptoms/signs	Pest and description	Management options	Remarks
Cottony white tufts of waxy material on needles. Yellow spots on needles are associated with cottony tufts.	COOLEY SPRUCE GALL APHID (ADELGID), *Adelges cooleyi.* Less than 1/16 inch long, purplish, immobile insects beneath cottony tufts. Suck sap.	Usually not a problem on landscape trees. If it is, apply carbaryl in spring.	Add a miticide to the spray tank to prevent a possible spider mite problem.

Table 7. *Continued*

Symptoms/signs	**Pest and description**	**Management options**	**Remarks**
	Elm		
Stickiness and blackening of foliage from honeydew and sooty mold. Cast skins on underside of leaves.	ELM LEAF APHID, *Tinocallis ulmifolii.* Green insects less than ⅛ inch long. Cluster on leaves. Suck sap.	Apply spray of diazinon. Inject trees with dicrotophos.	Inject only if spraying is not practical. Do not inject trees more than once each year.
Stickiness and blackening of foliage from honeydew and sooty mold. Tree may show decline and twig dieback.	EUROPEAN ELM SCALE, *Eriococcus spurius.* Dark red, immobile, oval insects about ⅛ inch long. Body is surrounded by white waxy fringe. Most common in twig crotches and undersides of limbs. Suck sap.	Apply oil in dormant season.	All elms are susceptible but Chinese elm supports highest populations. Insect has 1 generation each year.
Blackening of foliage from sooty mold. Tree may show decline and twig dieback.	EUROPEAN FRUIT LECANIUM, *Parthenolecanium corni.* Brown, flattened or hemispherical, immobile bodies, less than ¼ inch long, on twigs. Suck sap.	Normally held in check by natural enemies. If not, apply oil in dormant season. Or, apply carbaryl or diazinon when crawlers are numerous in late spring to early summer.	A greater problem in southern California than northern California. Insect has 1 generation each year.
Stickiness and blackening of foliage from honeydew and sooty mold. Stippling (flecking) of leaves. Leaves may appear bleached. Cast skins on underside of leaves.	LEAFHOPPERS: ROSE LEAFHOPPER, *Edwardsiana rosae; Empoasca* sp. Pale green to whitish, wedge-shaped insects up to ⅛ inch long. Suck sap.	Apply carbaryl.	In interior valleys add a miticide to the spray tank to prevent a possible spider mite problem.
Chewed leaves. Often single branches are defoliated.	SPINY ELM CATERPILLAR, *Nymphalis antiopa.* Larvae are black, covered with stiff black hairs, and 1½ inches long at maturity.	Ignore the caterpillars, or cut off branch tips containing a young colony.	
Skeletonized leaves. Entire tree often affected; leaves turn brown and fall.	ELM LEAF BEETLE, *Xanthogaleruca luteola.* Adults are yellow to olive green beetles, less than ¼ inch long, with several dark stripes down back. Larvae are yellow and black grubs, about ¼ inch long when mature. Chew leaves.	Apply spray of carbaryl when larvae of first generation begin feeding. In interior valleys, more than one spray each year is often needed. Or, inject trees with dicrotophos.	In interior valleys add a miticide to the spray tank to prevent a possible spider mite problem. Inject only if spraying is not possible or practical. Do not inject trees more than once each year.
Decline of branches or entire tree. Shotholes, about the size of pencil lead, in bark of affected woody parts.	SMALLER EUROPEAN ELM BARK BEETLE, *Scolytus multistriatus.* Adult beetles are about ⅛ inch long and brown to black. Larvae are white legless grubs, about ⅛ inch long when mature, found in galleries beneath bark. Adults transmit Dutch elm disease fungus. In presence or absence of disease, adults attack woody parts of stressed elms. Attack of adults and larvae can kill such trees.	Maintain elms in good state of vigor to discourage beetle attacks on woody parts. Do not prune elms when beetles are active (March to September). Elms showing yellowing or wilting in spring should be reported to County Agricultural Commissioner because of the possibility of Dutch elm disease. Get rid of declining or dead elm wood in which bettles breed.	Beetle has several generations each year. All species of elm are susceptible to beetle and to Dutch elm disease.

(continued)

Table 7. Elm *continued*

Symptoms/signs	Pest and description	Management options	Remarks
Holes, up to ½ inch in diameter, in trunk or limbs. Boring dust collects on bark plates or ground beneath holes. Tree may show decline.	CARPENTERWORM, *Prionoxystus robiniae*. When mature, larva is dirty white with brown head and 2½ inches long. Tunnels in cambium and heartwood.	To kill established borers, inject ethylene dichloride into gallery entrance holes, then plug hole entrance.	Insect has a 3 to 4 year life cycle. Adults emerge in spring and summer.

English Laurel

Symptoms/signs	Pest and description	Management options	Remarks
Decline or dieback of twigs, branches, or entire plant. Shotholes, about size of pencil lead, found in bark of affected plant parts.	SHOTHOLE BORER, *Scolytus rugulosus*. Adults are brown beetles less than ⅛ inch long. Larvae are white legless grubs, about ⅛ inch long when mature, and found in galleries beneath bark. Both stages will debilitate or kill already declining plants and occasionally healthy ones.	Keep plants in good state of vigor. Remove and destroy infested parts of plant.	
Decline or dieback of woody parts.	GREEDY SCALE, *Hemiberlesia rapax*. Immobile, grayish to brownish encrustations on twigs and branches. Individual scales are less than 1/16 inch long and circular to oval in shape. Suck sap, may inject toxic saliva into plant.	Various natural enemies normally prevent this scale from reaching damaging levels. If not, apply oil during dormant season. Or, apply diazinon when crawlers are numerous in spring and summer.	A heavy scale infestation is not likely to be reduced to low levels by a single oil application. Several consecutive years of annual oil sprays probably will be needed. Insect has several generations each year.
Stippling (flecking) or bleaching of leaves. Small varnishlike specks of excrement on underside of leaves.	GREENHOUSE THRIPS, *Heliothrips haemorrhoidalis*. Adults are black, slender, and about 1/16 inch long. Larvae are smaller and yellowish. Suck sap from underside of leaves.	Apply malathion, acephate, or insecticidal soap at onset of feeding, before foliage becomes bleached.	Insect has 5 to 7 generations annually.

Table 7. *Continued*

Symptoms/signs	Pest and description	Management options	Remarks
	Eucalyptus		
Dying or dead limbs or entire trees. Broad galleries beneath bark.	EUCALYPTUS LONGHORNED BORER, *Phoracantha semipunctata.* Adult beetles are ¾ to 1¼ inches long, reddish brown, with yellow markings on wing covers. Larvae, found beneath bark, are off-white and 1 to 1½ inches long when mature.	Keep trees in good state of vigor. Management is under investigation.	To date, insect occurs only in southern California. Insect has 2 and possibly 3 generations each year.
	Fir		
Chewed foliage. Tree may be defoliated.	TUSSOCK MOTHS: DOUGLAS FIR TUSSOCK MOTH, *Orgyia pseudotsugata;* RUSTY TUSSOCK MOTH, *O. antiqua.* Hairy caterpillars about 1 inch long when mature. Have prominent colorful or drab tufts of hair.	Apply acephate, carbaryl, or *Bacillus thuringiensis.*	
Top of tree, or entire tree, killed.	FIR ENGRAVER, *Scolytus ventralis.* Adults are brown beetles slightly over ⅛ inch long. Larvae are white legless grubs found in galleries beneath bark.	Maintain trees in good state of vigor to reduce likelihood of attack. No insecticides are registered for protection of firs against engraver attack.	
	Flowering Fruit Trees		
Decline or dieback of twigs and branches.	ARMORED SCALES: OYSTERSHELL SCALE, *Lepidosaphes ulmi;* GREEDY SCALE, *Hemiberlesia rapax;* SAN JOSE SCALE, *Quadraspidiotus perniciosus.* Immobile, grayish to brownish encrustations on twigs and branches. Individual scales are less than 1/16 inch long and circular to elongate in shape. Suck sap, may inject toxic saliva into plant.	Various natural enemies normally prevent these scales from reaching damaging levels. If not, apply oil during dormant season. Or, apply diazinon when crawlers are numerous in spring and summer.	A heavy scale infestation is not likely to be reduced to low levels by a single oil application. Several consecutive years of annual oil sprays probably will be needed. These scales have several generations each year.

(continued)

Table 7. Flowering Fruit Trees *continued*

Symptoms/signs	Pest and description	Management options	Remarks
Blackening of foliage from sooty mold. Possibly decline or dieback of twigs or branches.	SOFT (UNARMORED) SCALES: BLACK SCALE, *Saissetia oleae;* BROWN SOFT SCALE, *Coccus hesperidum;* KUNO SCALE, *Eulecanium kunoense.* Immature scales are orange, yellow, or brown, oval, flattened, immobile insects up to 1/16 inch long. Adult brown soft scale remains flattened, brown, immobile insect slightly larger than 1/8 inch. Adult black scale is brown to black, hemispherical in profile, with raised letter "H" on back. Adult kuno scale is beadlike and mahogany colored. Suck sap.	Scales (except kuno scale) normally controlled by natural enemies. Apply carbaryl or diazinon whenever crawlers are numerous.	Brown soft scale has 3 to 5 overlapping generations each year. Black scale has 2 generations each year along the coast and 1 annual generation inland. Kuno scale has 1 generation each year. It is a problem in Alameda and Contra Costa counties at present, but also occurs in Butte and Lake counties. Most likely to be seen on flowering peach, plum, cherry, apple, and pyracantha. Optimum date for spraying is about July 1.
Stickiness and blackening of foliage from honeydew and sooty mold. Curling of leaves. Cast skins on underside of leaves.	APHIDS (several spp.). Green to gray insects, less than 1/8 inch long, clustered on leaves or growing points. Suck sap.	Apply diazinon, acephate, or insecticidal soap.	
Woody swellings (galls) on branches and roots. Cottony waxy material on branches.	WOOLLY APPLE APHID, *Eriosoma lanigerum.* Reddish-bodied insects, about 1/16 inch long, often covered by cottony wax. Suck sap.	Apply diazinon or insecticidal soap.	Likeliest to occur on flowering apple.
Decline or dieback of twigs, branches, or entire tree. Shotholes, about size of pencil lead, found in bark of affected woody parts. Gumming of woody parts.	SHOTHOLE BORER, *Scolytus rugulosus.* Adults are brown beetles less than 1/8 inch long. Larvae are white legless grubs, about 1/8 inch long when mature, and found in galleries beneath bark. Both stages will debilitate or kill already declining plants and occasionally healthy ones.	Keep plants in good state of vigor. Remove and destroy infested parts of plant.	Likeliest to occur on flowering peach, plum, and cherry.
Chewed leaves. Single branches or entire tree may be defoliated.	CATERPILLARS, including: REDHUMPED CATERPILLAR, *Schizura concinna;* WESTERN TUSSOCK MOTH, *Orgyia vetusta;* FRUITTREE LEAFROLLER, *Archips argyrospilus;* TENT CATERPILLARS, *Malacosoma* spp. Some are 1½ inches long when mature. Some are hairy and others naked. Leafrollers tie leaves together with silk.	Apply *Bacillus thuringiensis* or carbaryl. Cut off branch terminal to remove redhumped caterpillar colony.	Redhumped caterpillar often attacks single branch terminal only.

Table 7. *Continued*

Symptoms/signs	Pest and description	Management options	Remarks
		Fuchsia	
Severe distortion and thickening of terminal growth, younger leaves, and blossoms. Plant stunted.	FUCHSIA GALL MITE, *Aculops fuchsiae.* Microscopic whitish wormlike mites only 1/75 inch long.	Grow resistant fuchsias (see p. 15). Prune off distorted growth whenever seen, then apply carbaryl or endosulfan twice, 2 to 3 weeks apart.	If carbaryl is used, add a miticide to the spray tank to prevent a possible spider mite problem.
Yellowing and drying up of leaves. Stickiness and blackening of foliage from sooty mold.	GREENHOUSE WHITEFLY, *Trialeurodes vaporariorum.* Adults are about 1/16 inch long with powdery white wings. Immature whiteflies are immobile, flattened, yellowish to greenish oval bodies, up to 1/16 inch long, on underside of leaves. Suck sap.	Apply acephate, resmethrin, or insecticidal soap several times at 4 to 6 day intervals.	
Stippling (flecking) or bleaching of leaves. Small varnishlike specks of excrement on underside of leaves.	GREENHOUSE THRIPS, *Heliothrips haemorrhoidalis.* Adults are black, slender, and about 1/16 inch long. Larvae are smaller and yellowish. Suck sap from underside of leaves.	Apply malathion, acephate, or insecticidal soap at onset of feeding, before foliage becomes bleached.	Insect has 5 to 7 generations annually.
Cupping of leaves. Stunted, twisted terminal growth. Blackening of foliage from sooty mold. Cast skins on underside of leaves.	POTATO APHID, *Macrosiphum euphorbiae.* Green to pink insects less than 1/8 inch long. Suck sap.	Apply acephate or insecticidal soap.	No thickening of plant tissue results from aphid infestation, as it does when fuchsia gall mite is present.
Blackening of foliage from sooty mold. Cottony waxy material on plant.	MEALYBUGS: CITRUS MEALYBUG, *Planococcus citri;* LONGTAILED MEALYBUG, *Pseudococcus longispinus.* Powdery gray insects, not more than 1/4 inch long when mature. Fringe of waxy filaments around edge of body. Suck sap.	Usually under biological control by natural enemies. If not, apply malathion or diazinon.	
Blackening of foliage or woody parts from sooty mold. Possibly decline of plant.	BLACK SCALE, *Saissetia oleae.* Adults are dark brown to black, immobile, bulbous insects, 1/8 to 3/16 inch long, with raised letter "H" on back. Immature scales are orange to brown, oval, flattened insects. Suck sap.	Often under adequate biological control by natural enemies. If not, apply carbaryl or diazinon when crawlers are numerous.	Insect has 2 generations each year along the coast and 1 annual generation inland.
Decline or dieback of twigs.	GREEDY SCALE, *Hemiberlesia rapax.* Immobile, grayish encrustations on woody parts. Individual scales are less than 1/16 inch long and nearly circular in shape. Suck sap, may inject toxic saliva into plant.	Various natural enemies normally prevent this scale from reaching damaging levels. If not, apply oil during dormant season. Or, apply diazinon when crawlers are numerous in spring and summer.	A heavy scale infestation is not likely to be reduced to low levels by a single oil application. Several consecutive years of annual oil sprays probably will be needed. Insect has several generations each year.

(continued)

Table 7. Fuchsia *continued*

Symptoms/signs	Pest and description	Management options	Remarks
Chewed leaves and blossoms.	FULLER ROSE BEETLE, *Pantomorus cervinus.* Adults are pale brown weevils with blunt snout and about ⅜ inch long.	Apply acephate.	
Stippling (flecking) of leaves. Fine webbing may be present. Leaves become bleached or reddened and may fall prematurely.	TWOSPOTTED SPIDER MITE, *Tetranychus urticae.* Tiny specks about the size of ground pepper, on leaves. Suck sap.	Apply dicofol or insecticidal soap several times at 7 to 10 day intervals.	
	Gardenia		
Blackening of foliage from sooty mold. Possibly decline of twigs or branches.	BROWN SOFT SCALE, *Coccus hesperidum.* Immature scales are yellow to brown, oval, flattened, immobile insects up to 1/16 inch long. Adult scale is flattened, brown, and slightly larger. Suck sap.	Normally controlled by natural enemies. If not, apply carbaryl whenever crawlers are numerous.	Scale has 3 to 5 overlapping generations each year.
Chewed leaves and blossoms.	FULLER ROSE BEETLE, *Pantomorus cervinus.* Adults are pale brown weevils with blunt snout and about ⅜ inch long.	Apply acephate.	
Blackening of foliage from sooty mold. Cottony waxy material on plant.	OBSCURE MEALYBUG, *Pseudococcus obscurus.* Grayish, heavily segmented insects lightly covered by powdery wax. Up to 3/16 inch long. Filaments at tail end are longer than those around sides of body. Suck sap.	Apply malathion.	
	Grecian Laurel		
Stippling (flecking) or bleaching of leaves. Small varnishlike specks of excrement on underside of leaves.	GREENHOUSE THRIPS, *Heliothrips haemorrhoidalis.* Adults are black, slender, and about 1/16 inch long. Larvae are smaller and yellowish. Suck sap from underside of leaves.	Apply malathion, acephate, or insecticidal soap at onset of feeding, before foliage becomes bleached.	Insect has 5 to 7 generations annually.

(continued)

Table 7. Grecian Laurel *continued*

Symptoms/signs	Pest and description	Management options	Remarks
Decline or dieback of twigs and branches.	OLEANDER SCALE, *Aspidiotus nerii.* Immobile, grayish encrustations on twigs, branches, or leaves. Individual scales are less than 1/16 inch long and circular to oval in shape. Suck sap, may inject toxic saliva into plant.	Various natural enemies normally prevent this scale from reaching damaging levels. If not, apply oil during dormant season. Or, apply diazinon when crawlers are numerous in spring and summer.	A heavy scale infestation is not likely to be reduced to low levels by a single oil application. Several consecutive years of annual oil sprays probably will be needed. Insect has several generations each year.
Margins of leaves rolled inward to form galls. Galls turn red, then brown.	LAUREL PSYLLID, *Trioza alacris.* Nymphs are about 1/16 inch long, covered with powdery substance, and found within galls. Suck sap.	Apply carbaryl or diazinon.	

Hebe

Symptoms/signs	Pest and description	Management options	Remarks
Blackening of foliage from sooty mold. White, immobile, popcornlike bodies on woody parts. Possibly decline of plant.	COTTONY CUSHION SCALE, *Icerya purchasi.* Adult females produce cottony white egg sacs. Immature scales are orange to brown, flattened, immobile insects on foliage or woody parts. Suck sap.	Normally under biological control by natural enemies. If not, apply carbaryl or diazinon when crawlers are numerous.	Insect breeds year-round.

Holly

Symptoms/signs	Pest and description	Management options	Remarks
Blackening of foliage from sooty mold. Possibly decline or dieback of twigs or branches.	SOFT (UNARMORED) SCALES: BROWN SOFT SCALE, *Coccus hesperidum;* BLACK SCALE, *Saissetia oleae.* Immature scales are yellow, orange, or brown oval, flattened, immobile insects less than 1/16 inch long. Adult brown soft scale is flattened, brown insect slightly more than 1/8 inch long. Adult black scale is brown or black, bulbous, with raised letter "H" on back. Suck sap.	Normally under adequate biological control by natural enemies. If not, apply carbaryl or diazinon when crawlers are active.	Brown soft scale has 3 to 5 overlapping generations each year. Black scale has 2 generations each year along the coast and 1 annual generation inland.

(continued)

Table 7. Holly *continued*

Symptoms/signs	Pest and description	Management options	Remarks
Decline or dieback of twigs or branches.	ARMORED SCALES: OYSTERSHELL SCALE, *Lepidosaphes ulmi;* GREEDY SCALE, *Hemiberlesia rapax;* OLEANDER SCALE, *Aspidiotus nerii.* Immobile, grayish to brownish encrustations on twigs and branches. Individual scales are less than 1/16 inch long and circular to elongate in shape. Suck sap, may inject toxic saliva into plant.	Various natural enemies normally prevent these scales from reaching damaging levels. If not, apply oil during dormant season. Or, apply diazinon when crawlers are numerous in spring and summer.	A heavy scale infestation is not likely to be reduced to low levels by a single oil application. Several consecutive years of annual oil sprays probably will be needed. These scales have several generations each year.
Slender, winding mines, or blotch mines, in leaves. Pinprick-like scars on leaves.	NATIVE HOLLY LEAFMINER, *Phytomyza ilicicola.* Adults are tiny black flies. Larvae are flattened, pale yellow insects found in mines in leaves.	Apply trichlorfon for adult control in April. Repeat in June.	Mines occur only in American holly. Puncture scars occur on American and Japanese hollies.

Honey Locust

Symptoms/signs	Pest and description	Management options	Remarks
Leaflets converted to podlike galls which turn brown and fall, leaving sections of branches without foliage.	HONEYLOCUST POD GALL MIDGE, *Dasineura gleditschiae.* Adult is a fly about 1/16 inch long. Larvae are white maggots, about 1/8 inch long when mature, found in leaf galls.	Where new plantings are planned, consider alternate trees. Insecticides do not effectively control this insect in landscape plantings.	Insect has 8 to 10 generations each year.
Tents of silk in tree. Leaves chewed.	MIMOSA WEBWORM, *Homadaula anisocentra.* Larvae are pale gray to dark brown with longitudinal white stripes. About 1/2 inch long when mature.	Prune off branch tips with webs when first seen and tents are small. Or, apply *Bacillus thuringiensis,* carbaryl, or acephate.	Most common in Sacramento Valley.

Hypericum

Symptoms/signs	Pest and description	Management options	Remarks
Stippling (flecking) or bleaching of leaves. Small varnishlike specks of excrement on underside of leaves.	GREENHOUSE THRIPS, *Heliothrips haemorrhoidalis.* Adults are black, slender, and about 1/16 inch long. Larvae are smaller and yellowish. Suck sap from underside of leaves.	Apply malathion, acephate, or insecticidal soap at onset of feeding, before foliage becomes bleached.	Insect has 5 to 7 generations annually.

(continued)

Table 7. Hypericum *continued*

Symptoms/signs	Pest and description	Management options	Remarks
Chewed leaves. Entire planting may be defoliated.	KLAMATHWEED BEETLE, *Chrysolina quadrigemina*. Adults are oval, metallic blue, and about ¼ inch long.	Apply carbaryl.	Damage occurs in spring only.

Iceplant

Symptoms/signs	Pest and description	Management options	Remarks
Decline of plants. Spots in planting die out. Cottony popcornlike material on plants. Blackening of plants from sooty mold.	ICEPLANT SCALES: *Pulvinaria delottoi, Pulvinariella mesembryanthemi*. Immature scales are emerald green, oval, and occur on leaves. Adult females are brown and produce cottony white egg sacs which look like popcorn. Suck sap.	Biological control program shows promise for population reduction statewide. As stopgap measure, apply oil plus malathion, diazinon, or carbaryl after peak of "popcorn" stage when crawlers are numerous.	New plantings should be started from cuttings known to be free of scales.
Wet, white, frothy masses of spittle on plants.	MEADOW SPITTLEBUG, *Philaenus spumarius*. Greenish bugs found within spittle masses. Suck sap.	No noticeable plant injury has been documented.	Common from April to June.

Ivy

Symptoms/signs	Pest and description	Management options	Remarks
Chewed leaves. Planting may be defoliated.	OMNIVOROUS LOOPER, *Sabulodes caberata*. Larvae are yellow, green, or pink with lengthwise stripes of yellow, green, or black. Crawl in "looping" manner. About 1½ inches long when mature.	Apply carbaryl, acephate, or *Bacillus thuringiensis*.	Sporadic pest which only occasionally reaches damaging numbers.
Poor growth or dieback of plants.	ARMORED SCALES: OLEANDER SCALE, *Aspidiotus nerii*; GREEDY SCALE, *Hemiberlesia rapax*; LATANIA SCALE, *H. lataniae*; YELLOW SCALE, *Aonidiella citrina*. Immobile, grayish to brownish encrustations on stems or foliage. Individual scales are less than 1/16 inch long and circular to oval in shape. Suck sap, may inject toxic saliva into plant.	Various natural enemies normally prevent these scales from reaching damaging levels. If not, apply oil during dormant season. Or, apply diazinon when crawlers are numerous in spring and summer.	A heavy scale infestation is not likely to be reduced to low levels by a single oil application. Several consecutive years of annual oil sprays probably will be needed. Removing leaves by mowing, before spraying, will improve coverage. These scales have several generations each year.

(continued)

Table 7. Ivy *continued*

Symptoms/signs	Pest and description	Management options	Remarks
Blackening of foliage from sooty mold.	SOFT (UNARMORED) SCALES: BROWN SOFT SCALE, *Coccus hesperidum;* NIGRA SCALE, *Parasaissetia nigra.* Immature scales are yellowish to brown, oval, flattened, immobile insects up to 1/16 inch long. Adults are slightly longer; brown soft scale is brown; nigra scale is black. Suck sap.	Normally controlled by natural enemies. If not, apply diazinon whenever crawlers are numerous.	Removing leaves by mowing, before spraying, will improve coverage. Brown soft scale has 3 to 5 overlapping generations each year; nigra scale has 1 generation annually.
Blackening of foliage from sooty mold. Cottony waxy material on plants.	GRAPE MEALYBUG, *Pseudococcus maritimus.* Powdery gray insects, not more than ¼ inch long when mature. Fringe of waxy filaments around edge of body. Suck sap.	Usually under biological control by natural enemies. If not, apply malathion or diazinon.	
Stickiness and blackening of plants from honeydew and sooty mold. Terminal growth sometimes stunted. Cast skins on underside of leaves.	APHIDS: GREEN PEACH APHID, *Myzus persicae;* IVY APHID, *Aphis hederae.* Green to nearly black insects less than ⅛ inch long. Cluster on leaves or terminals. Suck sap.	Apply acephate, diazinon, malathion, or insecticidal soap.	
Wet, white, frothy masses of spittle on plants.	MEADOW SPITTLEBUG, *Philaenus spumarius.* Greenish bugs found within spittle masses. Suck sap.	No noticeable plant injury has been documented.	Common from April to June.

Juniper

Symptoms/signs	Pest and description	Management options	Remarks
Scattered dying or dead branches. Entire plant is never dead.	JUNIPER TWIG GIRDLER, *Periploca nigra.* Larva is off-white with brown head and about ⅜ inch long when mature. Found in tunnel beneath bark. Girdles twigs.	Apply lindane twice: in southern California spray in late March and early May; in northern California spray in early June and mid-July. Pruning out affected branches improves appearance of planting but does not control girdler. Avoid planting Tam juniper.	Most common on Tam juniper. Add a miticide to the spray tank to prevent a possible spider mite problem. Insect has 1 generation each year.
Browning of tips of growth. Browning begins in fall and is at its worst in late winter to spring.	CYPRESS TIP MINER, *Argyresthia cupressella.* Adults are silvery tan moths with wingspan about ¼ inch. Larvae are green, about ¼ inch long when mature, and tunnel inside foliage.	Plant resistant junipers (see p. 14). Apply acephate, diazinon, or carbaryl when moths are active, usually March to April in southern California and April to May in northern California.	To determine when moths are active, shake foliage beginning in early March. Moths will fly out, then return to foliage. One annual generation. Insect occurs along coast only. If carbaryl is used, add a miticide to the spray tank to prevent a possible spider mite problem.

(continued)

Table 7. Juniper *continued*

Symptoms/signs	Pest and description	Management options	Remarks
Browning of tips of growth. Plant appears brown most of the year.	JUNIPER NEEDLE MINER, *Stenolechia bathrodyas.* Larvae are green, about ¼ inch long when mature, and mine leaflets.	Apply diazinon or acephate. Applications must coincide with each of the 3 annual flights of moths.	At this time, insect occurs only in southern California immediately along coast. May be confused with cypress tip miner.
Whitish to brownish encrustations on foliage. Possibly discolored foliage.	ARMORED SCALES: JUNIPER SCALE, *Carulaspis juniperi;* MINUTE CYPRESS SCALE, *C. minima.* Individual scales are circular to oval, immobile, and about 1/16 inch long. Suck sap.	Populations in California have very seldom warranted control. If control is necessary, apply diazinon when crawlers are numerous in May to June.	Insects have several generations each year.
Dieback of branches of Hollywood (Twisted Chinese) juniper.	FLATHEADED BORER, *Chrysobothris* sp. Larvae are white, horseshoe nail-shaped, and about 1¼ inches long when mature. Tunnel in woody parts.	Keep plants in good state of vigor.	Problem is most severe in San Joaquin Valley.
		Insecticidal controls not tested.	Other junipers may also be susceptible.

Laurel Fig (Indian Laurel)

Symptoms/signs	Pest and description	Management options	Remarks
Curling and purple pitting of terminal leaves.	CUBAN LAUREL THRIPS, *Gynaikothrips ficorum.* Adults are slender black insects, about ⅛ inch long, found among curled leaves.	Apply malathion, diazinon, or acephate to terminal foliage.	More of a problem in southern California than northern California. Insect has more than 5 annual generations. *Ficus microcarpa* is preferred host of this insect.

Lilac

Symptoms/signs	Pest and description	Management options	Remarks
Stunting and dieback of woody parts.	OYSTERSHELL SCALE, *Lepidosaphes ulmi.* Gray to brown encrustations on twigs and branches. Individual scales, about 1/16 inch long, look like miniature oysters. Suck sap, may inject toxic saliva into plant.	Various natural enemies normally prevent this scale from reaching damaging levels. If not, apply oil during dormant season. Or, apply diazinon when crawlers are numerous in spring and summer.	A heavy scale infestation is not likely to be reduced to low levels by a single oil application. Several consecutive years of annual oil sprays probably will be needed.
			Insect has 1 generation each year in northern California; 2 in southern California.

Table 7. *Continued*

Symptoms/signs	Pest and description	Management options	Remarks
		Liquidambar	
Webbing or tents on ends of branches. Chewed foliage.	FALL WEBWORM, *Hyphantria cunea.* Larvae are white to yellow, with long light-colored hairs, and about 1 inch long when mature. Found in colonies.	Cut off branch tip when tent first seen. Or, apply *Bacillus thuringiensis,* carbaryl, or acephate.	Most common in central San Joaquin Valley.
Chewed leaves. Entire tree may be defoliated. Tan, feltlike cocoons on trunk or limbs.	WESTERN TUSSOCK MOTH, *Orgyia vetusta.* Larvae have numerous red, blue, and yellow spots; long black tufts of hair at head and back end; and white tufts of hair on back. About 1 inch long when mature.	Presence of cocoons with egg masses attached is indicator of damage likely to occur. Scraping off cocoons before eggs attached hatch in spring is practical control on small trees. Apply *Bacillus thuringiensis,* carbaryl, or acephate.	Female is flightless and often lays eggs on her cocoon after she emerges as moth. Insect has 1 generation each year.
Chewed foliage. Single branches are often defoliated.	REDHUMPED CATERPILLAR, *Schizura concinna.* Larva is yellowish with lengthwise reddish, black, and red stripes, and with brick red head. About 1½ inches long when mature.	Cut off branch terminal to remove caterpillar colony. Or, apply *Bacillus thuringiensis,* carbaryl, or acephate.	
Blackening of foliage from sooty mold.	CALICO SCALE, *Eulecanium cerasorum.* Mature scales are black with prominent white or yellow spots, about ¼ inch long, globular, and immobile. Found on twigs. Suck sap.	Apply carbaryl or diazinon when crawlers are numerous. Or, apply oil in dormant season.	Most common in central California. Insect has 1 generation each year.
		Madrone	
Leaves with elliptical holes ⅛ to ¼ inch long.	MADRONE SHIELDBEARER, *Coptodisca arbutiella.* When young, larva makes winding mine in leaf. Mine turns brown. When mature, larva is about ¼ inch long and cuts hole in leaf.	Management never investigated.	
Blackening of foliage from sooty mold. Scalelike bodies on twigs.	MADRONE PSYLLID, *Euphyllura arbuti.* Adult is about ⅛ inch long, reddish, with membranous wings. Each nymph is covered by grayish immobile waxy cap. Suck sap.	Management never investigated.	Effect of feeding on tree is not known.

(continued)

Table 7. Madrone *continued*

Symptoms/signs	Pest and description	Management options	Remarks
Tents or mats of silk in tree. Chewed leaves.	WESTERN TENT CATERPILLAR, *Malacosoma californicum.* Larvae are predominantly brown, hairy, and nearly 2 inches long when mature.	At night or on rainy day when caterpillars are inside tent, cut off branch terminal with tent. Or, apply *Bacillus thuringiensis*, carbaryl, or acephate.	
Dark, immobile bodies, about 1/16 inch long, on underside of leaves.	MADRONE WHITEFLY, *Trialeurodes madroni.* Dark pupal cases have fringe of white filaments around all sides. Suck sap.	Management never investigated.	No damage to tree is recognized.

Magnolia

Symptoms/signs	Pest and description	Management options	Remarks
Decline or dieback of twigs or branches.	ARMORED SCALES: CALIFORNIA RED SCALE, *Aonidiella aurantii;* GREEDY SCALE, *Hemiberlesia rapax;* OLEANDER SCALE, *Aspidiotus nerii.* Immobile, grayish to brownish encrustations on twigs and branches. Individual scales are less than 1/16 inch long and circular to oval in shape. Suck sap, may inject toxic saliva into plant.	Various natural enemies normally prevent these scales from reaching damaging levels. If not, apply oil during dormant season. Or, apply diazinon when crawlers are numerous in spring and summer.	A heavy scale infestation is not likely to be reduced to low levels by a single oil application. Several consecutive years of oil sprays probably will be needed. These scales have several generations each year.
Blackening of foliage from sooty mold. White, immobile, popcornlike bodies on woody parts.	COTTONY CUSHION SCALE, *Icerya purchasi.* Adult females produce cottony white egg sacs. Immature scales are orange to brown, flattened, immobile insects on foliage or woody parts. Suck sap.	Normally under biological control by natural enemies. If not, apply carbaryl or diazinon when crawlers are numerous.	Insect breeds year-round.
Stippling (flecking) or bleaching of leaves. Small varnishlike specks of excrement on underside of leaves.	GREENHOUSE THRIPS, *Heliothrips haemorrhoidalis.* Adults are black, slender, and about 1/16 inch long. Larvae are smaller and yellowish. Suck sap from underside of leaves.	Apply malathion, acephate, or insecticidal soap at onset of feeding, before foliage becomes bleached.	Insect has 5 to 7 generations annually.

Table 7. *Continued*

Symptoms/signs	Pest and description	Management options	Remarks
	Mahonia (Oregon Grape)		
Chewed leaves. Plant may be defoliated.	BARBERRY LOOPER, *Coryphista meadii.* Olive green caterpillars, about 1 inch long when mature.	Apply carbaryl, acephate, or *Bacillus thuringiensis.*	Insect has several generations each year.
Decline or dieback of twigs or branches.	GREEDY SCALE, *Hemiberlesia rapax.* Immobile, grayish encrustations on woody parts. Individual scales are less than 1/16 inch long and nearly circular in shape. Suck sap, may inject toxic saliva into plant.	Various natural enemies normally prevent this scale from reaching damaging levels. If not, apply oil during dormant season. Or, apply diazinon when crawlers are numerous in spring and summer.	A heavy scale infestation is not likely to be reduced to low levels by a single oil application. Several consecutive years of annual oil sprays probably will be needed. Insect has several generations each year.
Poor plant growth.	DEER BRUSH WHITEFLY, *Aleurothrixus interrogationis.* Nymphs are yellow to tan, oval, immobile bodies, about 1/16 inch long, on underside of leaves. Pupae are nearly black, about same size. Suck sap.	Apply diazinon or acephate for control of nymphs.	
	Manzanita		
Leaves with elliptical holes about 1/8 to 1/4 inch long.	MADRONE SHIELD-BEARER, *Coptodisca arbutiella.* When young, larva makes winding mine in leaf. Mine turns brown. When mature, larva is about 1/4 inch long and cuts hole in leaf.	Management never investigated.	
Fleshy red galls on leaves. Reduction in new shoot growth.	MANZANITA LEAF GALL APHID, *Tamalia coweni.* Grayish insects, about 1/16 inch long, found in galls on leaves. Suck sap.	Ignore low levels of infestation. Apply diazinon or acephate.	
Stickiness and blackening of foliage from honeydew and sooty mold. Cast skins on plant. Shoots dying back.	APHID, *Wahlgreniella nervata.* Pink to green insects less than 1/8 inch long. Feed openly on leaves and terminals. Suck sap.	Apply acephate, diazinon, or insecticidal soap.	An early season problem.

(continued)

Table 7. Manzanita *continued*

Symptoms/signs	Pest and description	Management options	Remarks
Blackening of foliage from sooty mold. Possibly decline or dieback of twigs or branches.	BROWN SOFT SCALE, *Coccus hesperidum.* Immature scales are yellow to brown, oval, flattened, immobile insects up to 1/16 inch long. Adult is brown, flattened, immobile insect slightly longer than 1/8 inch. Suck sap.	Normally controlled by natural enemies. If not, apply carbaryl or diazinon whenever crawlers are numerous.	Brown soft scale has 3 to 5 overlapping generations each year.
Decline or dieback of twigs or branches.	ARMORED SCALES, including: MANZANITA SCALE, *Diaspis manzanitae;* OLEANDER SCALE, *Aspidiotus nerii;* GREEDY SCALE, *Hemiberlesia rapax.* Immobile, grayish to brownish encrustations on foliage or woody parts. Individual scales are less than 1/16 inch long and circular to oval in shape. Suck sap, may inject toxic saliva into plant.	Various natural enemies normally prevent these scales from reaching damaging levels. If not, apply oil during dormant season. Or, apply diazinon when crawlers are numerous in spring and summer.	A heavy scale infestation is not likely to be reduced to low levels by a single oil application. Several consecutive years of annual oil sprays probably will be needed. These scales have several generations each year.
Blackening of foliage from sooty mold. Immobile, dark, oval bodies about 1/16 inch long, on underside of leaves.	WHITEFLIES, including: CROWN WHITEFLY, *Aleuroplatus coronata;* IRIDESCENT WHITEFLY, *Aleuroparadoxus iridescens.* Immature whiteflies often have white waxy fringe around their bodies. Adults are powdery white flying insects about 1/16 inch long. Suck sap.	If blackening becomes objectionable, apply acephate or insecticidal soap.	
Blackening of foliage from sooty mold. Cottony waxy material on plant.	MEALYBUGS: *Puto arctostaphyli; P. albicans.* Powdery white, up to 1/4 inch long, with waxy fringe around edge of body. Occur on leaves and woody parts. Suck sap.	Management never investigated.	Most common above 2,000 feet elevation.
Chewed leaves. May be silk webbing on plant.	CATERPILLARS: WESTERN TUSSOCK MOTH, *Orgyia vetusta;* TENT CATERPILLAR, *Malacosoma* sp. Hairy, colorful or brown caterpillars 1 to 2 inches long when mature.	Apply *Bacillus thuringiensis,* carbaryl, or acephate.	
Dieback of branches or entire plant.	PACIFIC FLATHEADED BORER, *Chrysobothris mali.* Off-white, horseshoe nail-shaped larvae found in galleries in woody parts. About 3/4 inch long when mature.	Remove affected plant part. Keep plants in good state of vigor.	

Table 7. *Continued*

Symptoms/signs	Pest and description	Management options	Remarks
	Maple		
Stickiness and blackening of foliage from honeydew and sooty mold. Cast skins on underside of leaves. May be premature loss of leaves.	APHIDS, including: PAINTED MAPLE APHID, *Drepanaphis acerifolii; Periphyllus* spp. Green insects, 1/16 to 1/8 inch long, clustered on leaves. Suck sap.	Apply acephate, diazinon, or insecticidal soap as spray.	Insecticidal soap may injure Japanese maple.
Decline or dieback of twigs or branches.	ARMORED SCALES, including: OYSTERSHELL SCALE, *Lepidosaphes ulmi;* OLEANDER SCALE, *Aspidiotus nerii;* SAN JOSE SCALE, *Quadraspidiotus perniciosus.* Immobile, grayish to brownish encrustations on twigs and branches. Individual scales are less than 1/16 inch long and circular to elongate in shape. Suck sap, may inject toxic saliva into plant.	Various natural enemies normally prevent these scales from reaching damaging levels. If not, apply oil during dormant season. Or, apply diazinon when crawlers are numerous in spring and early summer.	A heavy scale infestation is not likely to be reduced to low levels by a single oil application. Several consecutive years of annual oil sprays probably will be needed. Do not apply oil to Japanese maple. These scales have several generations each year.
Stickiness and blackening of foliage from honeydew and sooty mold. Possibly some decline or dieback of woody parts.	SOFT (UNARMORED) SCALES, including: COTTONY MAPLE SCALE, *Pulvinaria innumerabilis;* CALICO SCALE, *Eulecanium cerasorum;* BLACK SCALE, *Saissetia oleae.* Immature scales are yellow to brown, oval, flattened, immobile insects up to 1/16 inch long. Adult cottony maple scale looks like popcorn. Adult calico scale is black with prominent white or yellow spots, and globular. Adult black scale is dark brown to black, bulbous, with raised letter "H" on back. Suck sap.	Normally controlled by natural enemies. If not, apply carbaryl or diazinon when crawlers are numerous.	Cottony maple scale and calico scale have 1 generation per year; crawlers are most numerous in late spring to early summer. Black scale has 2 generations each year along the coast and 1 annual generation inland. Cottony maple scale is most noticeable in mountainous and foothill areas of northern California; calico scale in central California.
Spotting or yellowing of foliage.	WESTERN BOXELDER BUG, *Leptocoris rubrolineatus.* Adults are 1/2 inch long and gray with prominent red diagonal markings. Suck sap.	Insecticidal controls not tested.	Insect is more important as an invading household pest than as a pest of landscape trees.

(continued)

Table 7. Maple *continued*

Symptoms/signs	Pest and description	Management options	Remarks
Dieback of woody parts.	FLATHEADED BORERS: FLATHEADED APPLE TREE BORER, *Chrysobothris femorata;* PACIFIC FLAT-HEADED BORER, *C. mali.* Larvae are dirty white and shaped like horseshoe nails. About ½ to 1 inch long at maturity. Mine beneath bark or into heartwood.	Take steps to return tree to a state of good vigor.	Insects attack weakened trees or trees in a declining state of health.

Mayten (Maytenus)

Symptoms/signs	Pest and description	Management options	Remarks
Blackening of foliage from sooty mold.	SOFT (UNARMORED) SCALES: BLACK SCALE, *Saissetia oleae;* NIGRA SCALE, *Parasaissetia nigra.* Immature scales are yellowish to brown, oval, flattened, immobile insects up to 1/16 inch long. Adult black scale is bulbous, brown to black, with raised letter "H" on back. Adult nigra scale is elongate and black. Suck sap.	Normally controlled by natural enemies. If not, apply carbaryl or diazinon whenever crawlers are numerous.	Black scale has 2 annual generations along coast and 1 inland. Nigra scale has 1 annual generation.

Mulberry

Symptoms/signs	Pest and description	Management options	Remarks
Webbing, or tents, on ends of branches. Chewed foliage.	FALL WEBWORM, *Hyphantria cunea.* Larvae are white to yellow, with long light-colored hairs, and about 1 inch long when mature. Found in colonies.	Cut off branch tip when tent first seen. Or, apply *Bacillus thuringiensis*, carbaryl, or acephate.	Most common in central San Joaquin Valley.
Decline or dieback of twigs or branches.	ARMORED SCALES: OLEANDER SCALE, *Aspidiotus nerii;* SAN JOSE SCALE, *Quadraspidiotus perniciosus;* CALIFORNIA RED SCALE, *Aonidiella aurantii.* Immobile, grayish to brownish encrustations on twigs and branches. Individual scales are less than 1/16 inch long and circular to oval in shape. Suck sap, may inject toxic saliva into plant.	Various natural enemies normally prevent these scales from reaching damaging levels. If not, apply oil during dormant season. Or, apply diazinon when crawlers are numerous in spring and summer.	A heavy scale infestation is not likely to be reduced to low levels by a single oil application. Several consecutive years of annual oil sprays probably will be needed. These scales have several generations each year.

Table 7. *Continued*

Symptoms/signs	Pest and description	Management options	Remarks
	Oak		
Chewed leaves. Tree may be defoliated.	CALIFORNIA OAKWORM, *Phryganidia californica*. Larvae are black with lengthwise yellow stripes, and are about 1¼ inches long when mature.	If larval numbers, leaf loss, or dropping fecal pellets early in larval feeding period indicate imminent damage, apply *Bacillus thuringiensis*, carbaryl, or acephate.	Primarily a coastal problem. Two generations each year in northern California; more, and sometimes overlapping, generations occur in southern California. Outbreaks requiring treatment do not occur every year.
			Spring oakworm generation is not usually important on deciduous oaks. Deciduous and live oaks are damaged by summer generation.
Chewed leaves. Leaves tied together with silk. Tree may be defoliated.	FRUITTREE LEAFROLLER, *Archips argyrospilus*. Larvae are green with black head and shield on back behind head. About ¾ inch long when mature. Wriggle vigorously when touched.	Apply carbaryl, diazinon, or *Bacillus thuringiensis*.	One generation each year. An early season problem. Most damage seen in San Bernardino Mts. and Pasadena areas of southern California, and Sierra foothills in northern California.
Leaves etched on one surface by larval feeding. These "windows" turn brown.	OAK RIBBED CASEMAKER, *Bucculatrix albertiella*. Larvae are about ¼ inch long when mature. More commonly seen are white, ribbed cigar-shaped cocoons on leaves.	Apply carbaryl.	Probably 2 generations each year, one in spring, the other in summer. Occurs on both deciduous and live oaks.
Chewed leaves. May or may not be silk tents or mats of silk in tree.	TENT CATERPILLARS: WESTERN TENT CATERPILLAR, *Malacosoma californicum;* PACIFIC TENT CATERPILLAR, *M. constrictum*. Larvae are hairy, mostly brown, with blue spots and orange or white tufts. Nearly 2 inches long at maturity.	Apply *Bacillus thuringiensis*, carbaryl, or acephate.	One generation each year. Larvae occur in spring.

(continued)

Table 7. Oak *continued*

Symptoms/signs	Pest and description	Management options	Remarks
Chewed leaves. Entire tree may be defoliated. Tan, feltlike cocoons on trunk or limbs.	WESTERN TUSSOCK MOTH, *Orgyia vetusta.* Larvae have numerous red, blue, and yellow spots; long black tufts of hair at head and back end; and white tufts on back. About 1 inch long when mature.	Presence of egg masses on cocoons is indicator of amount of damage likely to occur.	Female is flightless and often lays eggs on her cocoon after she emerges as a moth. Insect has 1 generation each year.
		Scraping off cocoons before attached eggs hatch in spring is practical control on small trees.	
		Apply *Bacillus thuringiensis*, carbaryl, or acephate.	
Gouging and etching of leaves. Leaves may turn brown.	LIVE OAK WEEVIL, *Deporaus glastinus.* Adults are dark metallic blue-green snout beetles, about ¼ inch long.	Apply carbaryl.	Most common on live oak but occurs occasionally on deciduous oaks. Most damage occurs from April to June.
Greatly roughened bark on lower trunk or crotches of major limbs. Retarded tree growth.	WESTERN SYCAMORE BORER, *Synanthedon resplendens.* Pink larva which tunnels in bark or cambium region. About ¾ inch long when mature.	Apply carbaryl to trunk and major crotches during adult emergence period (May through July).	Empty pupal cases protruding from roughened bark area indicate moth emergence. More prevalent in southern California than northern California.
Large holes, up to ½ inch in diameter, in trunks or limbs. Boring dust collects in bark plates or on ground below. Tree may show decline.	CARPENTERWORM, *Prionoxystus robiniae.* When mature, larva is dirty white with brown head and 2½ inches long. Tunnels in cambium and heartwood.	To kill established borers, inject ethylene dichloride into gallery entrance holes, then plug hole entrance.	Insect has a 3 to 4 year life cycle. Adults emerge in spring and summer.
Bleeding, or frothy material bubbling from tiny holes in trunk or limbs, often surrounded by a pile of fine boring dust.	OAK BARK BEETLES, *Pseudopityophthorus* spp. Adults are brown beetles 1/16 to ⅛ inch long. Larvae are white legless grubs found in galleries beneath bark.	Prevent attacks by maintaining tree in good state of vigor.	Presence of insect indicates that the tree is under stress.
Patches of dead leaves at ends of branches of live oaks.	OAK TWIG GIRDLER, *Agrilus angelicus.* Adults are whitish, look like links of sausage connected together, and are ¾ to 1 inch long when mature. Found beneath bark of affected twigs and branches, girdling spirally in cambium, or when mature, mining in heartwood.	Maintain trees in good state of vigor. As protectant, apply carbaryl in early June on coast, and in early May inland.	Most prevalent in southern California.
Swellings (galls) sometimes colorful, on leaves, flowers, twigs, or branches.	GALLS, formed by many species of cynipid wasps.	Most galls are not known to be harmful to the tree.	Heavy galling does not usually occur on the same tree year after year.

(continued)

Table 7. Oak *continued*

Symptoms/signs	Pest and description	Management options	Remarks
Blackened foliage from sooty mold. Twigs may have roughened bark and some may be killed.	OAK TREEHOPPER, *Platycotis vittata.* Adults are ¼ inch long, olive green to brown with red dots, and have thick horn on head. Nymphs and adults found on twigs, feeding in groups. Suck sap. Females chisel their eggs into twig bark.	Apply carbaryl.	
Dead twigs and branches. On deciduous oaks, dead leaves persist on tree in winter.	OAK PIT SCALE, *Asterolecanium minus.* Scales are pinhead-size, brown to olive green, and found on twigs. Scales often surrounded by doughnutlike swelling. Suck sap, may inject toxic saliva into plant.	Apply oil plus carbaryl or oil plus diazinon in late May or June. Or, apply oil alone in late dormant season, but before buds open.	In heavy infestations several annual applications will be necessary to reduce scale numbers to low level. More common on native deciduous oaks than on live oaks.
Blackening of foliage from sooty mold.	OAK LECANIUM SCALE, *Parthenolecanium quercifex.* Mature scales are brown, hemispherical in profile, and about ¼ inch long. They are immobile and occur on twigs. Suck sap.	Apply oil plus malathion for crawler control in late April to early May.	Greatest problem occurs on coast live oak in southern California. Insect has 1 generation each year.
Blackening of foliage from sooty mold. Immobile, dark, oval bodies about ¹⁄₁₆ inch long, on underside of leaves.	WHITEFLIES: CROWN WHITEFLY, *Aleuroplatus coronata;* GELATINOUS WHITEFLY, *A. gelatinosus;* STANFORD WHITEFLY, *Tetraleurodes stanfordi.* Immature whiteflies are immobile, oval, flattened, and often have white waxy fringe around their bodies. Adults are powdery white flying insects about ¹⁄₁₆ inch long. Suck sap.	Ignore immature insects for they are not known to injure vegetation. Where adults are a nuisance on trees close to home, apply malathion when adults emerge in spring.	Most common on live oaks.
Leaves with rolled margins.	WOOLLY OAK APHID, *Stegophylla quercicola.* Greenish to bluish insects, less than ⅛ inch long, covered with cottony wax fibers. Found within rolled leaf margins. Suck sap.	Management never investigated.	Insect occurs on coast live oak.
Raised blisters on leaves. Orange felty material fills depressions on underside of leaves.	COAST LIVE OAK ERINEUM MITE, *Aceria mackiei.* Minute wormlike mites not seen without magnification.	Management never investigated.	Mite occurs on coast live oak.

Table 7. *Continued*

Symptoms/signs	Pest and description	Management options	Remarks
Oleander			
Blackening of foliage from sooty mold.	BLACK SCALE, *Saissetia oleae.* Adults are immobile, dark brown to black bulbous insects, ⅛ to 3/16 inch long, with raised letter "H" on back. Immature scales are orange to brown, oval, flattened insects on woody parts and sometimes leaves. Suck sap.	Often under adequate biological control by natural enemies. If not, apply carbaryl or diazinon when crawlers are numerous. Thinning growth, permitting entry of more light and better air circulation, creates environment less favorable for scale.	Insect has 2 generations each year along the coast and 1 annual generation inland.
Stickiness and blackening of foliage from honeydew and sooty mold. Cast skins on plant.	OLEANDER APHID, *Aphis nerii.* Yellow and black insects, about 1/16 inch long. Cluster on growing points. Suck sap.	Often under adequate biological control by natural enemies. If not, apply acephate, diazinon, or insecticidal soap. Reduced watering and pruning will retard new shoot growth and prevent high aphid populations.	
Blackening of plant from sooty mold. Cottony waxy material on plant.	LONGTAILED MEALYBUG, *Pseudococcus longispinus.* Powdery white insects about ⅛ inch long when mature. Filaments at tail end are as long or longer than body. Suck sap.	Normally under biological control by natural enemies. If not, apply diazinon or malathion.	
Olive			
Blackening of foliage from sooty mold. Possibly reduced growth of tree.	BLACK SCALE, *Saissetia oleae.* Adults are immobile, dark brown to black bulbous insects, ⅛ to 3/16 inch long, with raised letter "H" on back. Immature scales are orange to brown, oval, flattened insects. Occur on woody parts and sometimes leaves. Suck sap.	Often under adequate biological control by natural enemies. If not, apply carbaryl or diazinon when crawlers are numerous. Thinning tree growth, permitting entry of more light and better air circulation, creates environment less favorable for scale.	Insect has 2 generations each year along the coast and 1 annual generation inland.

(continued)

Table 7. Olive *continued*

Symptoms/signs	Pest and description	Management options	Remarks
Decline or dieback of twigs and branches.	ARMORED SCALES, including: OLEANDER SCALE, *Aspidiotus nerii;* OLIVE SCALE, *Parlatoria oleae.* Immobile, grayish to brownish encrustations on twigs and branches. Individual scales are less than 1/16 inch long and circular to oval in shape. Suck sap, may inject toxic saliva into plant.	Various natural enemies normally prevent these scales from reaching damaging levels. If not, apply oil during dormant season. Or, apply diazinon when crawlers are numerous in spring and summer.	A heavy scale infestation is not likely to be reduced to low levels by a single oil application. Several consecutive years of annual oil sprays probably will be needed. These scales have several generations each year.
Dieback of occasional twigs.	BRANCH AND TWIG BORER, *Polycaon confertus.* Adults are ¼ to ½ inch long, brown to black beetles. Tunnel into twigs.	Prune off affected plant parts. Eliminate beetle breeding sources close to olive trees.	Beetles breed in dead oak, maple, bay, eucalyptus, and other hardwoods.

Palm

Symptoms/signs	Pest and description	Management options	Remarks
Yellowing or dieback of fronds.	ARMORED SCALES, including: GREEDY SCALE, *Hemiberlesia rapax;* OLEANDER SCALE, *Aspidiotus nerii;* CALIFORNIA RED SCALE, *Aonidiella aurantii;* BOISDUVAL SCALE, *Diaspis boisduvalii.* Immobile, grayish to brownish encrustations on twigs and branches. Individual scales are less than 1/16 inch long and circular to oval in shape. Suck sap, may inject toxic saliva into sap.	Various natural enemies normally prevent these scales from reaching damaging levels. If not, apply oil during dormant season. Or, apply diazinon when crawlers are numerous in spring and summer.	A heavy scale infestation is not likely to be reduced to low levels by a single oil application. Several consecutive years of annual oil sprays probably will be needed. Oil may injure some species of palm. These scales have several generations each year.
Blackening of foliage from sooty mold. Possibly yellowing and decline of fronds.	SOFT (UNARMORED) SCALES: BROWN SOFT SCALE, *Coccus hesperidum;* BLACK SCALE, *Saissetia oleae;* HEMISPHERICAL SCALE, *S. coffeae.* Immature scales are yellow to brown, oval, flattened, immobile insects up to 1/16 inch long. Adult brown soft scale remains flattened, brown insect. Adult black scale is dark brown to black, bulbous, with raised letter "H" on back. Adult hemispherical scale is dark brown, bulbous, without letter "H." Suck sap.	Normally controlled by natural enemies. If not, apply carbaryl or diazinon whenever crawlers are numerous.	Brown soft scale has 3 to 5 overlapping generations each year. Black scale has 2 generations each year along the coast and 1 inland.

(continued)

Table 7. Palm *continued*

Symptoms/signs	Pest and description	Management options	Remarks
Blackening of plant from sooty mold. Cottony waxy material on plant.	MEALYBUGS: LONGTAILED MEALYBUG; *Pseudococcus longispinus;* OBSCURE MEALYBUG, *P. obscurus.* Powdery white insects up to 3/16 inch long when mature. Fringe of waxy filaments around edge of body. Those at tail end are longest. Suck sap.	Normally under biological control by natural enemies. If not, apply malathion or diazinon.	
Stippling (flecking) or bleaching of foliage. Small varnishlike specks of excrement on underside of leaves.	GREENHOUSE THRIPS, *Heliothrips haemorrhoidalis.* Adults are black, slender, and about 1/16 inch long. Larvae are smaller and yellowish. Suck sap from underside of leaves.	Apply malathion, acephate, or insecticidal soap at onset of feeding before foliage becomes bleached.	Insect has 5 to 7 generations annually.

Pepper

Symptoms/signs	Pest and description	Management options	Remarks
Doughnutlike pits in leaflets, and in petioles and young twigs. Trees appear grayish green and sparsely foliated.	PEPPERTREE PSYLLID, *Calophya schini.* Adults are greenish and about 1/16 inch long. Nymphs are immobile and develop in pits. Suck sap.	Apply acephate.	Insect appears to reproduce year-round along the coast; repeated spray applications about 60 days apart may be necessary. A pest of California pepper tree. Brazilian pepper, *Schinus terebinthifolius,* is resistant.
Blackening of foliage from sooty mold. Possibly decline or dieback of twigs and branches.	SOFT (UNARMORED) SCALES: BLACK SCALE, *Saissetia oleae;* HEMISPHERICAL SCALE, *S. coffeae;* BARNACLE SCALE, *Ceroplastes cirripediformis.* Immature scales are yellow to brown, oval, flattened, immobile insects up to 1/16 inch long. Adult black scale is dark brown to black, bulbous, with raised letter "H" on back. Adult hemispherical scale is brown, without letter "H." Adult barnacle scale is gray and resembles tiny barnacle. Suck sap.	Normally controlled by natural enemies. If not, apply carbaryl or diazinon whenever crawlers are numerous.	Black scale has 2 generations each year along the coast and 1 inland.

(continued)

Table 7. Pepper *continued*

Symptoms/signs	Pest and description	Management options	Remarks
Decline or dieback of twigs or branches.	ARMORED SCALES: GREEDY SCALE, *Hemiberlesia rapax;* OLEANDER SCALE, *Aspidiotus nerii.* Immobile, grayish to brownish encrustations on twigs and branches. Individual scales are less than 1/16 inch long and circular to oval in shape. Suck sap, may inject toxic saliva into plant.	Various natural enemies normally prevent these scales from reaching damaging levels. If not, apply oil during dormant season. Or, apply diazinon when crawlers are numerous in spring and summer.	A heavy scale infestation is not likely to be reduced to low levels by a single oil application. Several consecutive years of annual oil sprays probably will be needed. These scales have several generations each year.
Leaves chewed.	OMNIVOROUS LOOPER, *Sabulodes caberata.* Larvae are yellow, green, or pink with lengthwise stripes of yellow, green, or black. Crawl in "looping" manner. About 1½ inches long when mature.	Apply carbaryl, acephate, or *Bacillus thuringiensis.*	Sporadic pest which only occasionally reaches damaging numbers.

Photinia

Symptoms/signs	Pest and description	Management options	Remarks
Stippling (flecking), bleaching, or reddening (bronzing) of leaves. Small varnishlike specks of excrement on underside of leaves.	GREENHOUSE THRIPS, *Heliothrips haemorrhoidalis.* Adults are black, slender, and about 1/16 inch long. Larvae are smaller and yellowish. Suck sap from underside of leaves.	Apply malathion, acephate, or insecticidal soap at onset of feeding, before foliage becomes bleached.	Insect has 5 to 7 generations annually.

Pine

Symptoms/signs	Pest and description	Management options	Remarks
Stickiness, varnishing, or blackening of foliage from honeydew and sooty mold. Possibly yellowing of needles.	APHIDS: *Essigella californica; Schizolachnus piniradiatae.* Green to gray slender insects, about 1/16 inch long, on needles. Suck sap.	Apply acephate, diazinon, or insecticidal soap.	A more severe problem in southern California than northern California.
Blackening of foliage from sooty mold. Possibly yellowing of older needles.	IRREGULAR PINE SCALE, *Toumeyella pinicola.* When mature, females are about ¼ inch long. Resemble immobile chips of marble on 1 to 2 year old twig growth. Males resemble grains of rice and occur on needles. Suck sap.	Apply carbaryl or diazinon twice (May, June) during crawler emergence period.	If carbaryl is used, add a miticide to the spray tank to prevent a possible spider mite problem. Insect has 1 generation each year.

(continued)

Table 7. Pine *continued*

Symptoms/signs	Pest and description	Management options	Remarks
Yellow mottling, or dieback, of needles.	ARMORED SCALES: PINE NEEDLE SCALE, *Chionaspis pinifoliae;* BLACK PINE LEAF SCALE, *Nuculaspis californica.* White to gray or black immobile bodies on needles. About 1/16 inch long. Suck sap.	Usually under biological control by natural enemies. If not, apply diazinon when crawlers are numerous.	Both scales have several generations each year in southern California and 1 in northern California.
Cottony white or grayish material on bark of trunk, limbs, twigs, or needles. Tree may show poor growth.	ADELGIDS (PINE BARK APHIDS), *Pineus* spp. Purplish insects, less than 1/16 inch long, beneath cottony material. Suck sap.	Apply carbaryl in spring.	Add a miticide to the spray tank to prevent a possible spider mite problem. Addition of extra wetting agent, and high pressure, will improve control.
Stippling (flecking) of needles. Entire tree may become bleached. More common on young trees.	SPIDER MITES: *Oligonychus subnudus; O. milleri.* Greenish to pink specks, about the size of finely ground pepper, on needles. Suck sap.	First, determine presence of increasing numbers of mites by periodic shaking of foliage over a white surface. Apply dicofol or fenbutatin-oxide.	*O. subnudus* is most common along coast; causes most damage in spring to early summer. *O. milleri* is most destructive inland, during summer to fall.
Needles chewed and notched along their length. Affected needles turn brown. Damage seen in late winter, early spring.	PINE NEEDLE WEEVILS, *Scythropus* spp. Adults are brown with blunt snout, and nearly 1/4 inch long. Occur on needles.	Management never investigated.	Few needles are affected and these soon drop. Needles of the preceding year's growth only are attacked. Larvae feed on roots but no effect on tree is recognized.
Mined buds and shoot tips. Killed tips give tree red or brown appearance. Tree becomes bunchy.	PINE TIP MOTHS: NANTUCKET PINE TIP MOTH, *Rhyacionia frustrana;* MONTEREY PINE TIP MOTH, *R. pasadenana;* PONDEROSA PINE TIP MOTH, *R. zozana.* Larvae are orange to brown, occur in mined buds and shoot tips, and are about 5/8 inch long when mature.	Apply acephate or dimethoate when adults are flying and laying eggs, but before larvae enter plant tissue. Nantucket pine tip moth has 3 to 4 annual generations. Spray May 1 and repeat every 5 weeks until September 1. Other species have 1 annual generation with moths flying in spring; spray twice during that period. Infestations by Nantucket pine tip moth can be prevented or reduced by planting resistant pines (see p. 15).	Nantucket pine tip moth at present occurs in southern California; Monterey pine tip moth along entire coast; ponderosa pine tip moth in higher elevations of inland central and northern California. Nantucket pine tip moth is under good control in some southern California localities.
Tips of Monterey pine mined for distance of 1 to 2 inches only. Tips die, often in crooked position.	MONTEREY PINE BUD MOTH, *Exoteleia burkei.* Larvae are brownish yellow and 3/16 inch long when mature. Found in mined pine tips.	Management never investigated.	Very little tissue is affected.
Pitchy masses 1 to 4 inches in diameter protruding from trunk and limbs. Breakage of limbs weakened by insect may occasionally be seen.	SEQUOIA PITCH MOTH, *Synanthedon sequoiae.* Larvae are dirty white with brown head, and about 1 inch long when mature. Found beneath pitch mass, in cambium region.	Avoid pruning or otherwise injuring woody parts from February to September, for adult insects are attracted to resin bleeding from wounds. Scrape off pitch masses and kill larvae.	Satisfactory control has not been achieved by sprays.

(continued)

Table 7. Pine *continued*

Symptoms/signs	Pest and description	Management options	Remarks
Trees dead, sometimes dying quickly. Boring dust found on bark plates or collecting in crotches. Galleries seen beneath bark of affected trees.	BARK BEETLES (SCOLYTIDAE). Adults are brown to nearly black beetles, ⅛ to nearly ¼ inch long (except red turpentine beetle). Larvae are white legless grubs about the same size at maturity. Both stages found beneath bark of infested trees.	Keep pines in good state of vigor, for attack often begins on stressed pines. Protect specimen pines with preventive spray of lindane or carbaryl.	Identification of beetles is needed to distinguish aggressive species from more benign ones, and to make proper management decision. The most destructive one, the California fivespined ips, has up to 4 generations each year. A spray applied about mid-February will protect pines that entire season. Sprays made later will protect only against attack of later generations. If trees are sprayed with lindane or carbaryl, add a miticide to the spray tank to prevent a possible spider mite problem.
Tree may be declining or possibly dead. Pink or whitish pitch tubes, ¾ to 1 inch in diameter, protruding from bark at base of tree trunk only. Or, piles of pink or whitish coarse granular material in piles on ground at base of tree.	RED TURPENTINE BEETLE, *Dendroctonus valens*. Adults are cinnamon brown bark beetles about ¼ inch long. Larvae are white legless grubs, about ⅜ inch long when mature. Both stages found beneath bark at base of tree only, in brood galleries or "patches."	Keep pines in a good state of vigor, for insect cannot survive in a vigorous tree. If more than 1 pitch tube (or pile of granular material) is found per foot of trunk circumference, apply protective spray of lindane or carbaryl to tree base.	Beetle rarely attacks higher than 6 to 8 feet on tree trunk. Most attacks occur within lowest 2 feet of trunk. Insect is not usually a killer of trees, but may stress them further and invite attack by a more aggressive species of bark beetle. Beetle has up to 4 generations each year. A spray applied about mid-February will protect pines that entire season. Sprays made later will protect only against attack of later generations.
Tree may be declining or possibly dead.	FLATHEADED AND ROUNDHEADED BORERS (BUPRESTIDAE and CERAMBYCIDAE). Off-white, legless larvae found in brood galleries beneath bark. Some reach about 1 inch when mature. Flatheaded borer larvae have horseshoe nail shape.	Keep pines in a good state of vigor, for borers prefer stressed trees. Insecticidal sprays are not especially useful for control of these insects.	These borers may further stress trees and invite attack by bark beetles.
Wet, white, frothy masses of spittle on twigs or cones.	SPITTLEBUGS, *Aphrophora* spp. Greenish to blackish insects are found in spittle masses. Suck sap.	Injury to pines has not been documented.	
Chewed foliage.	SAWFLIES, *Neodiprion* spp. Greenish caterpillars are ¾ to 1 inch long at maturity. Found on foliage.	Apply carbaryl.	Add a miticide to the spray tank to prevent a possible spider mite problem.

(continued)

Table 7. Pine *continued*

Symptoms/signs	Pest and description	Management options	Remarks
Sections of shoots with greatly shortened needles with swollen, bulbous bases.	MONTEREY PINE MIDGE, *Thecodiplosis piniradiatae.* Tiny white legless grubs are found inside swollen needle base.	Management never investigated.	

Pittosporum

Symptoms/signs	Pest and description	Management options	Remarks
Thickening of sections of twigs. Shoots may be distorted or killed.	PIT-MAKING PITTOSPORUM SCALE, *Asterolecanium arabidis.* Mature scales are brown to white ⅛ inch long immobile insects on twigs. Suck sap.	Management never investigated.	An occasional problem only in northern California.
Stickiness and blackening of plant from honeydew and sooty mold. Cast skins on plant.	APPLE APHID, *Aphis pomi.* Bright green insects, about 1/16 inch long, clustered on growing points or leaves. Suck sap.	Often under biological control by natural enemies. If not, apply acephate, diazinon, or insecticidal soap.	
Blackening of foliage from sooty mold. White, immobile, popcornlike bodies on woody parts.	COTTONY CUSHION SCALE, *Icerya purchasi.* Adult females produce cottony white egg sacs. Immature scales are orange to brown, flattened, immobile insects on foliage or woody parts. Suck sap.	Normally under biological control by natural enemies. If not, apply carbaryl or diazinon when crawlers are numerous.	Insect breeds year-round.
Decline or dieback of twigs or branches.	GREEDY SCALE, *Hemiberlesia rapax.* Immobile, grayish encrustations on woody parts. Individual scales are less than 1/16 inch long and nearly circular in shape. Suck sap, may inject toxic saliva into plant.	Various natural enemies normally prevent this scale from reaching damaging levels. If not, apply oil during dormant season. Or, apply diazinon when crawlers are numerous in spring and summer.	A heavy scale infestation is not likely to be reduced to low levels by a single oil application. Several consecutive years of annual oil sprays probably will be needed. Insect has several generations each year.

Podocarpus

Symptoms/signs	Pest and description	Management options	Remarks
Bluish white bloom covering foliage. Cast skins on foliage.	PODOCARPUS APHID, *Neophyllaphis podocarpi.* Grayish insects, about 1/16 inch long, clustered on stems or leaves. Suck sap.	Apply acephate, diazinon, or insecticidal soap.	

(continued)

Table 7. Podocarpus *continued*

Symptoms/signs	Pest and description	Management options	Remarks
Decline or dieback of twigs or branches.	CALIFORNIA RED SCALE, *Aonidiella aurantii.* Immobile, brownish encrustations on woody parts. Individual scales are less than 1/16 inch long and circular to oval in shape. Suck sap, may inject toxic saliva into plant.	Normally under biological control by natural enemies. If not, apply diazinon or oil in the fall.	Insect has several generations each year.

Poplar (including Cottonwood and Aspen)

Symptoms/signs	Pest and description	Management options	Remarks
Chewed leaves. Tree may be defoliated. Leaves may be tied together with silk. May be silken tents in tree.	CATERPILLARS, including: FALL WEBWORM, *Hyphantria cunea;* REDHUMPED CATERPILLAR, *Schizura concinna;* SPOTTED HALISIDOTA, *Halisidota maculata;* FRUITTREE LEAFROLLER, *Archips argyrospilus.* Naked or hairy larvae, ¾ to 1½ inches long when mature.	Apply *Bacillus thuringiensis* or carbaryl.	
Stickiness or blackening of foliage from honeydew and sooty mold. Cast skins on underside of leaves.	APHIDS, including: CLOUDYWINGED COTTONWOOD APHID, *Periphyllus populicola; Chaitophorus* spp. Green to grayish insects less than ⅛ inch long. Cluster on leaves. Suck sap.	Apply diazinon or insecticidal soap.	
Blackening of foliage from sooty mold. Possibly some decline or dieback of woody parts.	SOFT (UNARMORED) SCALES, including: BROWN SOFT SCALE, *Coccus hesperidum;* COTTONY MAPLE SCALE, *Pulvinaria innumerabilis;* EUROPEAN FRUIT LECANIUM, *Parthenolecanium corni;* BLACK SCALE, *Saissetia oleae.* Immature scales are yellowish to brown, oval, flattened insects less than 1/16 inch long. Adult brown soft scale remains flattened and brown. Adult cottony maple scale looks like popcorn. Both adult European fruit lecanium and black scales are dark and bulbous, but black scale has raised letter "H" on back. Suck sap.	Normally under biological control by natural enemies. If not, apply carbaryl or diazinon when crawlers are numerous.	Brown soft scale has 3 to 5 overlapping generations each year. Black scale has 2 generations each year along the coast and 1 inland. Other scales have 1 generation each year.

(continued)

Table 7. Poplar *continued*

Symptoms/signs	Pest and description	Management options	Remarks
Decline or dieback of twigs or branches.	ARMORED SCALES, including: OYSTERSHELL SCALE, *Lepidosaphes ulmi;* SAN JOSE SCALE, *Quadraspidiotus perniciosus.* Immobile, grayish to brownish encrustations on twigs and branches. Individual scales are less than 1/16 inch long and circular to elongate in shape. Suck sap, may inject toxic saliva into plant.	Various natural enemies normally prevent these scales from reaching damaging levels. If not, apply oil during dormant season. Or, apply diazinon when crawlers are numerous in spring and summer.	A heavy scale infestation is not likely to be reduced to low levels by a single oil application. Several consecutive years of annual oil sprays probably will be needed. Oystershell scale has 2 generations each year in southern California and 1 in northern California. San Jose scale has several generations each year.
Warty, woody swellings (galls) on twigs.	COTTONWOOD GALL MITE, *Eriophyes parapopuli.* Wormlike mites are microscopic.	Management methods not tested.	
Dieback of branches or sometimes entire tree.	BORERS, including: FLATHEADED BORERS (BUPRESTIDAE); ROUNDHEADED BORERS (CERAMBYCIDAE); CLEARWING MOTHS (SESIIDAE); CARPENTERWORM, *Prionoxystus robiniae.* Larvae are off-white, from ¾ to 2½ inches long when mature, and mine beneath bark or in heartwood.	Keep trees in good state of vigor. Cut off and destroy dying branches whenever seen. To kill established carpenterworm or clearwing moth larvae, inject ethylene dichloride into entrance holes, then plug entrance.	Most borers attack declining trees or tree parts, and hasten the death of the tree or its branches.
Skeletonized leaves.	LEAF AND FLEA BEETLES: COTTONWOOD LEAF BEETLE, *Chrysomela scripta;* ASPEN LEAF BEETLE, *C. crotchi; Altica bimarginata.* Adults are brown to metallic black, oval beetles ¼ to ⅜ inch long. Larvae are dark in color and elongate.	Apply carbaryl.	

Privet

Symptoms/signs	Pest and description	Management options	Remarks
Blackening of foliage from sooty mold.	BLACK SCALE, *Saissetia oleae.* Adults are immobile, dark brown to black bulbous insects, ⅛ to 3/16 inch long, with raised letter "H" on back. Immature scales are orange to brown, oval, flattened insects. Suck sap.	Often under adequate biological control by natural enemies. If not, apply carbaryl or diazinon when crawlers are numerous.	Insect has 2 generations each year along the coast and 1 annual generation inland.

(continued)

Table 7. Privet *continued*

Symptoms/signs	Pest and description	Management options	Remarks
Decline or dieback of twigs or branches.	ARMORED SCALES: SAN JOSE SCALE, *Quadraspidiotus perniciosus;* CALIFORNIA RED SCALE, *Aonidiella aurantii.* Immobile, grayish to brownish encrustations on twigs and branches. Individual scales are less than 1/16 inch long and circular to oval in shape. Suck sap, may inject toxic saliva into plant.	Various natural enemies normally prevent these scales from reaching damaging levels. If not, apply oil during dormant season. Or, apply diazinon when crawlers are numerous in spring and summer.	A heavy scale infestation is not likely to be reduced to low levels by a single oil application. Several consecutive years of annual oil sprays probably will be needed. Scales have several generations each year.

Pyracantha

Symptoms/signs	Pest and description	Management options	Remarks
Stickiness and blackening of plant from honeydew and sooty mold. Cast skins on foliage.	APPLE APHID, *Aphis pomi.* Bright green insects, about 1/16 inch long, clustered on growing points or leaves. Suck sap.	Often under biological control by natural enemies. If not, apply acephate, diazinon, or insecticidal soap.	
Woody swellings (galls) on branches and roots. Cottony waxy material on branches. Plant may show poor growth.	WOOLLY APPLE APHID, *Eriosoma lanigerum.* Reddish-bodied insects, about 1/16 inch long, often covered by cottony wax. Suck sap.	Apply diazinon or insecticidal soap.	
Decline or dieback of twigs or branches.	ARMORED SCALES: GREEDY SCALE, *Hemiberlesia rapax;* SAN JOSE SCALE, *Quadraspidiotus perniciosus.* Immobile, grayish to brownish encrustations on twigs and branches. Individual scales are less than 1/16 inch long and circular to oval in shape. Suck sap, may inject toxic saliva into plant.	Various natural enemies normally prevent these scales from reaching damaging levels. If not, apply oil during dormant season. Or, apply diazinon when crawlers are numerous in spring and summer.	A heavy scale infestation is not likely to be reduced to low levels by a single oil application. Several consecutive years of annual oil sprays probably will be needed. Scales have several generations each year.
Chewed leaves. Plant may be defoliated.	WESTERN TUSSOCK MOTH, *Orgyia vetusta.* Larvae have spots of red and yellow, brushlike tufts of hair, and are 1 inch long when mature.	Apply *Bacillus thuringiensis,* carbaryl, or acephate.	
Stickiness and blackening of plant from honeydew and sooty mold. Plant may show poor growth.	KUNO SCALE, *Eulecanium kunoense.* Mature female scales are mahogany colored, beadlike, 1/8 to 3/16 inch long, and occur on twigs and branches. Suck sap.	Apply carbaryl after peak of crawler emergence has passed (about July 1).	A problem in Alameda and Contra Costa counties at present. Insect also occurs in Lake and Butte counties.

(continued)

Table 7. Pyracantha *continued*

Symptoms/signs	Pest and description	Management options	Remarks
Reddening, or bronzing, of foliage.	SPIDER MITE, *Oligonychus platani*. Greenish or brownish specks about the size of ground pepper. Occur on both leaf surfaces. Suck sap.	Apply dicofol or insecticidal soap.	More important in interior valleys than coastally.

Redbud (Cercis)

Symptoms/signs	Pest and description	Management options	Remarks
Decline or dieback of twigs or branches.	ARMORED SCALES: GREEDY SCALE, *Hemiberlesia rapax;* OLEANDER SCALE, *Aspidiotus nerii.* Immobile, grayish to brownish encrustations on twigs and branches. Individual scales are less than 1/16 inch long and circular to oval in shape. Suck sap, may inject toxic saliva into plant.	Various natural enemies normally prevent these scales from reaching damaging levels. If not, apply oil during dormant season. Or apply diazinon when crawlers are numerous in spring and summer.	A heavy scale infestation is not likely to be reduced to low levels by a single oil application. Several consecutive years of annual oil sprays probably will be needed. Scales have several generations each year.
Blackening of foliage from sooty mold.	GREENHOUSE WHITEFLY, *Trialeurodes vaporariorum.* Adults are about 1/16 inch long with powdery white wings. Immatures are immobile, flattened, yellowish to greenish oval bodies, up to 1/16 inch long, on undersides of leaves. Suck sap.	Spray with acephate, resmethrin, or insecticidal soap several times at 4 to 6 day intervals.	
Chewed leaves. May be silk tents or mats of silk on plant.	WESTERN TENT CATERPILLAR, *Malacosoma californicum.* Hairy brown caterpillars, nearly 2 inches long when mature.	At night or on rainy day when caterpillars are inside tent, cut off branch terminal with tent. Or apply *Bacillus thuringiensis*, carbaryl, or acephate.	

Redwood

Symptoms/signs	Pest and description	Management options	Remarks
Stippling (flecking) of needles. Tree loses normal color and appears lighter green or grayish.	SPRUCE SPIDER MITE, *Oligonychus ununguis.* Greenish specks about the size of ground pepper. Fine webbing often present on needles. Suck sap from needles.	First, determine presence of increasing mite numbers by periodic shaking of foliage over a white surface. If required, apply dicofol or fenbutatin-oxide.	A "cool weather" mite. Highest populations occur in spring and fall.

(continued)

Table 7. Redwood *continued*

Symptoms/signs	Pest and description	Management options	Remarks
Browning of tips of growth. Browning begins in fall and is at its worst in late winter to spring.	CYPRESS TIP MINER, *Argyresthia cupressella.* Adults are silvery tan moths with wingspan about ¼ inch. Larvae are green, about ¼ inch long when mature, and tunnel inside foliage.	Ignore the problem, for very few tips are likely to be mined.	
Grayish or brownish encrustations on needles or shoots. Infested needles may show yellowing.	ARMORED SCALES: REDWOOD SCALE, *Aonidia shastae;* BLACK ARAUCARIA SCALE, *Lindingaspis rossi.* Individual scales are circular or oval, immobile, and about 1/16 inch long. Suck sap.	Normally under biological control by natural enemies. If not, apply diazinon when crawlers are numerous in spring.	More of a problem in southern California than northern California.

Rhododendron

Symptoms/signs	Pest and description	Management options	Remarks
Wilting and death of plants. Some roots may be missing or their bark removed or plant may be girdled just below soil surface.	BLACK VINE WEEVIL, *Otiorhynchus sulcatus.* Adults are black snout beetles, nearly ½ inch long. Larvae occur in soil and are white legless grubs with brown head.	Apply acephate to plants and soil surface beginning when "scalloping" or notching of leaf margins occurs on rhododendron, photinia, pyracantha, and others. Sprays are intended to kill adult weevils before they lay the eggs that become the more destructive larval stage.	Scalloping of leaf margins begins about April. If adult control is not achieved, be advised that control of larvae in soil is extremely difficult. If possible, spray before or after bloom, for sprays may injure open blossoms.
		Plant resistant rhododendrons (see p. 16).	
Stippling (flecking) or bleaching of leaves. Small varnishlike specks of excrement on underside of leaves.	GREENHOUSE THRIPS, *Heliothrips haemorrhoidalis.* Adults are black, slender, and about 1/16 inch long. Larvae are smaller and yellowish. Suck sap from underside of leaves.	Apply malathion, acephate, or insecticidal soap at onset of feeding, before foliage becomes bleached.	Insect has 5 to 7 generations annually. Confine sprays to periods before or after bloom, for sprays may injure open blossoms.

Table 7. *Continued*

Symptoms/signs	Pest and description	Management options	Remarks
		Rose	
Stippling (flecking) of leaves. Leaves become bleached, dry up, and drop.	SPIDER MITES, *Tetranychus* spp. Greenish to pink specks about the size of ground pepper. Suck sap.	Apply dicofol, fenbutatin-oxide, or insecticidal soap. Several applications 7 to 10 days apart will be necessary.	Good underleaf coverage is necessary.
Blossom petals streaked with brown.	WESTERN FLOWER THRIPS, *Frankliniella occidentalis.* Adults are straw colored, slender insects less than 1/16 inch long. Found in blossoms. Rasp tissue, suck sap.	Apply acephate or diazinon.	
Stippling (flecking) of leaves. Stippled spots are larger than those caused by spider mites. Cast skins on underside of leaves. Leaves may appear bleached.	ROSE LEAFHOPPER, *Edwardsiana rosae.* Pale green to whitish, wedge-shaped insects up to 1/8 inch long. Suck sap.	Apply carbaryl.	Add a miticide to the spray tank to prevent a possible spider mite problem.
Stickiness and blackening of foliage from honeydew and sooty mold. Blossom distortion. Cast skins on plant.	APHIDS, including: ROSE APHID, *Macrosiphum rosae;* COTTON (MELON) APHID, *Aphis gossypii.* Green to pinkish insects, 1/16 to 1/8 inch long, clustered on growing points and flower buds. Suck sap.	Apply acephate, diazinon, or insecticidal soap.	
Blackening of foliage from sooty mold. Possibly decline of canes.	SOFT (UNARMORED) SCALES: BROWN SOFT SCALE, *Coccus hesperidum;* EUROPEAN FRUIT LECANIUM, *Parthenolecanium corni;* BLACK SCALE, *Saissetia oleae.* Immature scales are orange, yellow, or brown, oval, flattened, immobile insects up to 1/16 inch long. Adult brown soft scale remains flattened, immobile insect slightly longer than 1/8 inch. Adult European fruit lecanium is brown, bulbous insect. Adult black scale is similar but has raised letter "H" on back. Suck sap.	Normally controlled by natural enemies. If not, apply carbaryl or diazinon whenever crawlers are numerous.	Brown soft scale has 3 to 5 overlapping generations each year. European fruit lecanium has 1 generation each year. Black scale has 2 generations each year along the coast and 1 annual generation inland. If carbaryl is used, add a miticide to the spray tank to prevent a possible spider mite problem.
Decline or dieback of canes. Grayish or whitish encrustations on canes.	ARMORED SCALES, including: ROSE SCALE, *Aulacaspis rosae;* SAN JOSE SCALE, *Quadraspidiotus perniciosus;* GREEDY SCALE, *Hemiberlesia rapax.* Immobile, grayish to brownish encrustations on canes. Individual scales are less than 1/16 inch long and circular to oval in shape. Suck sap, may inject toxic saliva into plant.	Various natural enemies normally prevent these scales from reaching damaging levels. If not, apply oil during dormant season. Or, apply diazinon when crawlers are numerous in spring and summer.	A heavy scale infestation is not likely to be reduced to low levels by a single oil application. Several consecutive years of annual oil sprays probably will be needed. Scales have several generations each year.

(continued)

Table 7. Rose *continued*

Symptoms/signs	Pest and description	Management options	Remarks
Blackening of foliage from sooty mold. Yellowing of leaves.	GREENHOUSE WHITEFLY, *Trialeurodes vaporariorum.* Adults are about 1/16 inch long with powdery white wings. Immatures are immobile, flattened, yellowish to greenish oval bodies, up to 1/16 inch long, on underside of leaves. Suck sap.	Apply acephate, resmethrin, or insecticidal soap several times at 4 to 6 day intervals.	
Decline or death of canes or entire plant.	FLATHEADED BORERS: PACIFIC FLATHEADED BORER, *Chrysobothris mali;* FLATHEADED APPLE TREE BORER, *C. femorata.* Larvae are off-white, shaped like horseshoe nails, and about 1 inch long when mature. Tunnel in base of older canes or at plant crown.	Remove cane stubs left from earlier pruning. Keep plants in good state of vigor. Apply carbaryl as protective spray to plant crowns in mid to late spring.	Add a miticide to the spray tank to prevent a possible spider mite problem.
Chewed leaves and blossoms.	FULLER ROSE BEETLE, *Pantomorus cervinus.* Adults are pale brown weevils with blunt snout and about 3/8 inch long.	Apply acephate.	
Holes punched in flower buds. Open flowers riddled with holes and ragged in appearance. Holes gouged in green canes.	ROSE CURCULIO, *Rhynchites bicolor.* Adult is red to black weevil with prominent long snout. Insect is about 1/4 inch long.	Apply carbaryl.	Add a miticide to the spray tank to prevent a possible spider mite problem. Insect seems to prefer yellow and white rose varieties.
Tips of new canes wilt in spring. Canes may suffer dieback in summer.	RASPBERRY HORNTAIL, *Hartigia cressoni.* Segmented white larva about 1 inch long when mature. Found spirally girdling tips of canes. Or, found in central portion (pith) of cane or in roots.	Remove and destroy infested canes whenever found. If a problem in past years, apply carbaryl April 1 and repeat May 1.	A more severe problem in interior valleys than coastally. Add a miticide to the spray tank to prevent a possible spider mite problem.
Chewed leaves. Leaves may be tied together with silk.	CATERPILLARS AND LEAFROLLERS, including: FOREST TENT CATERPILLAR, *Malacosoma disstria;* OMNIVOROUS LOOPER, *Sabulodes caberata;* FRUITTREE LEAFROLLER, *Archips argyrospilus;* ORANGE TORTRIX, *Argyrotaenia citrana.* Hairy or naked larvae, 3/4 to 1 1/2 inches long when mature.	Apply acephate, carbaryl, or *Bacillus thuringiensis.*	If carbaryl is used, add a miticide to the spray tank to prevent a possible spider mite problem.
Leaves skeletonized. Later, large holes are eaten from leaves.	BRISTLY ROSESLUG, *Cladius difformis.* Pale green larva with bristlelike hairs. About 5/8 inch long when mature.	Apply carbaryl or diazinon.	Insect has 5 to 6 generations each year. If carbaryl is used, add a miticide to the spray tank to prevent a possible spider mite problem.

(continued)

Table 7. Rose *continued*

Symptoms/signs	Pest and description	Management options	Remarks
Semicircular holes in margins of leaves and flower petals.	LEAFCUTTER BEES, *Megachile* spp. Insect itself is seldom noticed.	Bees are beneficial and should not be controlled by sprays. No effective nonchemical means of management are known.	Bees line their cells (nests) with the plant material cut.

Rosemary

Symptoms/signs	Pest and description	Management options	Remarks
Wet, white, frothy masses of spittle on plants.	MEADOW SPITTLEBUG, *Philaenus spumarius.* Greenish bugs found within spittle masses. Suck sap.	No noticeable plant injury has been documented.	Common from April to June.

Sequoia

Symptoms/signs	Pest and description	Management options	Remarks
Stippling (flecking) of foliage. Tree loses normal color and appears lighter green or yellowish.	SPRUCE SPIDER MITE, *Oligonychus ununguis.* Greenish specks about the size of ground pepper. Fine webbing often present on needles. Suck sap from needles.	First, determine presence of increasing mite numbers by periodic shaking of foliage over a white surface. If necessary, apply dicofol or fenbutatin-oxide.	A "cool weather" mite. Highest populations occur in spring and fall.
Grayish or brownish encrustations on foliage. Infested foliage may show yellowing.	REDWOOD SCALE, *Aonidia shastae.* Individual scales are circular to oval, immobile, and about 1/16 inch long. Suck sap.	Normally under biological control by natural enemies. If not, apply diazinon when crawlers are abundant in spring.	

Spiraea

Symptoms/signs	Pest and description	Management options	Remarks
Foliage blackened from sooty mold. Leaves curled. Cast skins on foliage.	SPIREA APHID, *Aphis citricola.* Light- to dark-green insects, less than 1/8 inch long, clustered on growing points and leaves. Suck sap.	Apply diazinon, acephate, or insecticidal soap.	

(continued)

Table 7. Spiraea *continued*

Symptoms/signs	Pest and description	Management options	Remarks
Decline or dieback of twigs or branches.	ARMORED SCALES: OYSTERSHELL SCALE, *Lepidosaphes ulmi;* SAN JOSE SCALE, *Quadraspidiotus perniciosus.* Immobile, grayish to brownish encrustations on woody parts. Individual scales are less than 1/16 inch long and circular to elongate in shape. Suck sap, may inject toxic saliva into plant.	Various natural enemies normally prevent these scales from reaching damaging levels. If not, apply oil during dormant season. Or, apply diazinon when crawlers are numerous in spring and summer.	A heavy scale infestation is not likely to be reduced to low levels by a single oil application. Several consecutive years of annual oil sprays probably will be needed. Oystershell scale has 2 generations each year in southern California; 1 in northern California. San Jose scale has several generations each year.

Spruce

Symptoms/signs	Pest and description	Management options	Remarks
Yellowing of interior needles. Needles then drop, leaving only young needles at tips of branches.	SPRUCE APHID, *Elatobium abietinum.* Green insects about 1/16 inch long. Suck sap.	Apply diazinon, acephate, or insecticidal soap.	Insect attack occurs in late winter to very early spring. Insects are difficult to see because they blend with foliage color.
Stippling (flecking) of needles. Tree loses normal color and appears lighter green or yellowish.	SPRUCE SPIDER MITE, *Oligonychus ununguis.* Greenish specks about the size of ground pepper. Fine webbing often present on needles. Suck sap from needles.	First, determine presence of increasing mite numbers by periodic shaking of foliage over a white surface. If necessary, apply dicofol or fenbutatin-oxide.	A "cool weather" mite. Highest populations occur in spring and fall.
Pinecone-like galls on tips of branches. These later turn brown and become unsightly.	COOLEY SPRUCE GALL APHID (ADELGID), *Adelges cooleyi.* Adelgids responsible for galls are seldom seen.	Apply carbaryl in April to May, but only if galls have been seen on tree in a previous year.	Seen on spruce only in mountainous northern California and north coast. Add a miticide to the spray tank to prevent a possible spider mite problem.
Yellow mottling of needles.	PINE NEEDLE SCALE, *Chionaspis pinifoliae.* White, immobile bodies, about 1/16 inch long, on needles. Suck sap.	Usually under biological control by natural enemies. If not, apply diazinon in late spring when crawlers are numerous.	Insect has 2 generations each year in southern California; 1 in northern California.

Table 7. Continued

Symptoms/signs	Pest and description	Management options	Remarks
	Strawberry Tree		
Decline or dieback of twigs or branches.	GREEDY SCALE, *Hemiberlesia rapax*. Immobile, grayish encrustations on woody parts. Individual scales are less than 1/16 inch long and nearly circular in shape. Suck sap, may inject toxic saliva into plant.	Various natural enemies normally prevent this scale from reaching damaging levels. If not, apply oil during growing season. Or, apply diazinon when crawlers are numerous in spring and summer.	A heavy scale infestation is not likely to be reduced to low levels by a single oil application. Several consecutive years of annual oil sprays probably will be needed. Insect has several generations each year.
Blackening of foliage from sooty mold. Possibly decline or dieback of twigs or branches.	SOFT (UNARMORED) SCALES: BROWN SOFT SCALE, *Coccus hesperidum;* BLACK SCALE, *Saissetia oleae*. Immature scales are orange, yellow, or brown, oval, flattened, insects up to 1/16 inch long. Adult brown soft scale remains flattened, brown, immobile insect slightly larger than 1/8 inch. Adult black scale is dark brown to black, bulbous, with raised letter "H" on back. Suck sap.	Normally controlled by natural enemies. If not, apply carbaryl or diazinon whenever crawlers are numerous.	Brown soft scale has 3 to 5 overlapping generations each year. Black scale has 2 generations each year along the coast and 1 annual generation inland.
	Sycamore (Plane Tree)		
Yellow spots on leaves. Spots turn brown later in season and many leaves drop prematurely. In winter, cottony material protruding from bark crevices and from beneath bark plates on trunk and limbs.	SYCAMORE SCALE, *Stomacoccus platani*. Bulbous insects less than 1/16 inch long, found in center of yellow spots on lower surface of leaves. Suck sap.	Apply oil in January.	High pressure is needed for good control. Insect has 3 to 5 generations each year.
Stippling (flecking) of leaves. Leaves may become bleached.	SYCAMORE SPIDER MITE, *Oligonychus platani*. Green specks about the size of ground pepper. Suck sap.	Apply dicofol or fenbutatin-oxide.	Most damaging in interior valleys.
Stunting and dieback of twigs or branches.	OYSTERSHELL SCALE, *Lepidosaphes ulmi*. Gray to brown immobile encrustations on twigs and branches. Individual scales, about 1/16 inch long, look like minature oysters. Suck sap, may inject toxic saliva into plant.	Various natural enemies normally prevent this scale from reaching damaging levels. If not, apply oil during dormant season. Or, apply diazinon when crawlers are numerous in spring and summer.	A heavy scale infestation is not likely to be reduced to low levels by a single oil application. Several consecutive years of annual oil sprays probably will be needed. Insect has 2 generations each year in southern California; 1 in northern California.

(continued)

Table 7. Sycamore *continued*

Symptoms/signs	Pest and description	Management options	Remarks
Stippling (flecking) of leaves. Leaves may become bleached. Varnish-like specks of excrement, and cast skins, found on underside of leaves.	WESTERN SYCAMORE LACE BUG, *Corythucha confraterna.* Adults have lacy wings and are about ⅛ inch long. Nymphs are smaller and without wings. Both stages occur on underside of leaves. Suck sap.	Apply carbaryl.	More common in southern California than northern California.
Young leaves skeletonized, or holes seen in leaves.	SYCAMORE LEAF SKELETONIZER, *Gelechia desiliens.* Larvae are greenish, found in tubular nest of leaf hairs on leaves, and are ½ inch long when mature.	Apply *Bacillus thuringiensis,* carbaryl, or acephate.	
Greatly roughened bark on lower trunk or crotches of major limbs. Boring dust seen on roughened bark. Retarded tree growth.	WESTERN SYCAMORE BORER, *Synanthedon resplendens.* Pink larva which tunnels in bark or cambium region. About ¾ inch long when mature.	Apply carbaryl to trunk and major crotches during adult emergence period (May through July).	Empty pupal cases protruding from roughened bark area indicate moth emergence. More prevalent in southern California than northern California.

Tamarisk

Symptoms/signs	Pest and description	Management options	Remarks
Stunting or dieback of woody parts.	OYSTERSHELL SCALE, *Lepidosaphes ulmi.* Gray to brown immobile encrustations on twigs and branches. Individual scales, about 1/16 inch long, look like miniature oysters. Suck sap, may inject toxic saliva into plant.	Various natural enemies normally prevent this scale from reaching damaging levels. If not, apply oil during dormant season. Or, apply diazinon when crawlers are numerous in spring and summer.	A heavy scale infestation is not likely to be reduced to low levels by a single oil application. Several consecutive years of annual oil sprays probably will be needed. Insect has 2 generations each year in southern California; 1 in northern California.

Taxus (Yew)

Symptoms/signs	Pest and description	Management options	Remarks
Blackening of foliage from sooty mold. Cottony waxy material on plant.	OBSCURE MEALYBUG, *Pseudococcus obscurus.* Grayish, heavily segmented insects lightly covered by powdery wax. Up to 3/16 inch long. Filaments at tail end are longer than those around sides of body. Suck sap.	Apply malathion.	

(continued)

Table 7. Taxus *continued*

Symptoms/signs	Pest and description	Management options	Remarks
Decline or dieback of twigs or branches.	ARMORED SCALES: OLEANDER SCALE, *Aspidiotus nerii;* PURPLE SCALE, *Lepidosaphes beckii.* Immobile, grayish to brownish encrustations on twigs and branches. Individual scales are less than 1/16 inch long and circular to elongate in shape. Suck sap, may inject toxic saliva into plant.	Various natural enemies normally prevent these scales from reaching damaging levels. If not, apply oil during dormant season. Or, apply diazinon when crawlers are numerous in spring and summer.	A heavy scale infestation is not likely to be reduced to low levels by a single oil application. Several consecutive years of annual oil sprays probably will be needed. Scales have several generations each year.

Toyon (Christmas Berry)

Symptoms/signs	Pest and description	Management options	Remarks
Terminal leaves severely curled and twisted.	TOYON THRIPS, *Rhyncothrips ilex.* Adults are black, slender insects about 1/16 inch long. Larvae are smaller and orange-yellow. All stages found in growing points. Suck sap.	Apply acephate.	An early season problem. Insect has only 1 annual generation. Plants with bare soil beneath appear to sustain less damage.
Stippling (flecking) or bleaching of leaves. Varnishlike specks of excrement on underside of leaves.	GREENHOUSE THRIPS, *Heliothrips haemorrhoidalis.* Adults are black, slender, and about 1/16 inch long. Larvae are smaller and yellowish. Suck sap from underside of leaves.	Apply malathion, acephate, or insecticidal soap at onset of feeding, before foliage becomes bleached.	Insect has 5 to 7 generations annually.
Stippling (flecking) of leaves. Foliage bleached. Varnishlike specks of excrement and cast skins on underside of leaves.	LACE BUGS, *Corythucha* spp. Adults are brownish with lacelike wings and 1/8 to 1/4 inch long. Suck sap.	Apply carbaryl.	More serious in inland locations than coastally. Several generations each year.
Chewed leaves. May be silk tents or mats of silk in plant.	CATERPILLARS: WESTERN TENT CATERPILLAR, *Malacosoma californicum;* WESTERN TUSSOCK MOTH, *Orgyia vetusta.* Hairy caterpillars. May have colored spots. Reach 1 to 2 inches at maturity.	Apply *Bacillus thuringiensis,* carbaryl, or acephate.	One generation each year. Larvae occur in spring.
Blackening of foliage from sooty mold. Immobile, dark or light oval bodies, about 1/16 inch long, on underside of leaves.	WHITEFLIES, including: CROWN WHITEFLY, *Aleuroplatus coronatus;* IRIDESCENT WHITEFLY, *Aleuroparadoxus iridescens.* Immature whiteflies are immobile, oval, flattened, and often have white waxy fringe around their bodies. Adults are powdery white flying insects about 1/16 inch long. Suck sap.	If blackening of foliage is objectionable, apply acephate, resmethrin, or insecticidal soap several times at 4 to 6 day intervals.	

(continued)

Table 7. Toyon *continued*

Symptoms/signs	Pest and description	Management options	Remarks
Blackening of foliage from sooty mold. Possibly some decline or dieback of woody parts.	EUROPEAN FRUIT LECANIUM, *Parthenolecanium corni.* Brown, flattened, or hemispherical immobile bodies, less than ¼ inch long, on twigs. Suck sap.	Normally held in check by natural enemies. If not, apply oil in dormant season. Or, apply carbaryl or diazinon when crawlers are numerous in late spring to early summer.	Insect has 1 generation each year.
Dieback of branches or entire plant.	PACIFIC FLATHEADED BORER, *Chrysobothris mali.* Off-white, horseshoe nail-shaped larvae found in woody parts. About ¾ inch long when mature.	Remove affected plant part. Keep plants in good state of vigor.	

Tulip Tree (Liriodendron)

Symptoms/signs	Pest and description	Management options	Remarks
Stickiness and blackening of tree from honeydew and sooty mold. Cast skins on underside of leaves. Possibly premature yellowing of leaves.	TULIPTREE APHID, *Macrosiphum liriodendri.* Green insects less than ⅛ inch long, in colonies on undersides of leaves. Suck sap.	Apply acephate, diazinon, or insecticidal soap.	

Viburnum

Symptoms/signs	Pest and description	Management options	Remarks
Stippling (flecking) or bleaching of leaves. Small varnishlike specks of excrement on underside of leaves.	GREENHOUSE THRIPS, *Heliothrips haemorrhoidalis.* Adults are black, slender, and about 1/16 inch long. Larvae are smaller and yellowish. Suck sap from underside of leaves.	Apply malathion, acephate, or insecticidal soap at onset of feeding, before foliage becomes bleached.	Insect has 5 to 7 generations annually.
Stunting or dieback of woody parts.	OYSTERSHELL SCALE, *Lepidosaphes ulmi.* Gray to brown immobile encrustations on twigs and branches. Individual scales, about 1/16 inch long, look like miniature oysters. Suck sap, may inject toxic saliva into plant.	Various natural enemies normally prevent this scale from reaching damaging levels. If not, apply oil during dormant season. Or, apply diazinon when crawlers are numerous in spring and summer.	A heavy scale infestation is not likely to be reduced to low levels by a single oil application. Several consecutive years of annual oil sprays probably will be needed. Insect has 2 generations each year in southern California; 1 in northern California.

Table 7. *Continued*

Symptoms/signs	Pest and description	Management options	Remarks
	Walnut		
Stickiness or blackening of foliage from honeydew and sooty mold. Cast skins on underside of leaves.	APHIDS: WALNUT APHID, *Chromaphis juglandicola;* DUSKYVEINED APHID, *Callaphis juglandis.* Yellowish to brownish insects, up to ⅛ inch long. Found in colonies on either leaf surface. Suck sap.	Normally under biological control by natural enemies. If not, apply diazinon or insecticidal soap.	
Decline or dieback of twigs or branches.	ARMORED SCALES: WALNUT SCALE, *Quadraspidiotus juglansregiae;* SAN JOSE SCALE, *Q. perniciosus;* ITALIAN PEAR SCALE, *Epidiaspis leperii;* OYSTERSHELL SCALE, *Lepidosaphes ulmi.* Tan to gray immobile encrustations on twigs and branches. Individual scales are circular to elongate and up to 1/16 inch long. Suck sap, may inject toxic saliva into plant.	Apply oil-diazinon combination after leaves begin to show. Or, apply diazinon for crawler control in spring and summer.	Oil in dormant season may cause injury to walnut. Oystershell scale has 2 generations each year in southern California; 1 in northern California. Walnut scale and San Jose scale have several generations each year. Italian pear scale has 1 generation each year.
Blackening of foliage from sooty mold. Possibly some decline or dieback of woody parts.	SOFT (UNARMORED) SCALES: FROSTED SCALE, *Parthenolecanium pruinosum;* EUROPEAN FRUIT LECANIUM, *P. corni;* CALICO SCALE, *Eulecanium cerasorum.* Immature scales are yellowish to brown, immobile, flattened, oval bodies less than 1/16 inch long. Adults reach about ¼ inch. Adult frosted and European fruit lecanium scales are brown and bulbous. Adult calico scale is black and bulbous with prominent white or yellow spots. Suck sap.	Normally held in check by natural enemies. If not, apply carbaryl or diazinon when crawlers are numerous in late spring to early summer.	Scales have 1 generation each year.
Stippling (flecking) of leaves. May become bleached or reddened (bronzed).	SPIDER MITES: TWOSPOTTED SPIDER MITE, *Tetranychus urticae;* PACIFIC SPIDER MITE, *T. pacificus;* EUROPEAN RED MITE, *Panonychus ulmi.* Yellowish to reddish specks about the size of ground pepper. Suck sap.	Often under biological control by natural enemies. If not, apply dicofol.	

(continued)

Table 7. Walnut *continued*

Symptoms/signs	Pest and description	Management options	Remarks
Chewed leaves. Silken tents may occur on ends of branches. Single branches or entire tree may be defoliated.	CATERPILLARS: RED-HUMPED CATERPILLAR, *Schizura concinna;* FALL WEBWORM, *Hyphantria cunea.* About 1 to 1½ inches long when mature.	Cut off branch terminal to remove caterpillar colony. Or, apply *Bacillus thuringiensis* or carbaryl.	

Willow

Symptoms/signs	Pest and description	Management options	Remarks
Stickiness and blackening of foliage from honeydew and sooty mold.	APHIDS, including: GIANT WILLOW APHID, *Lachnus salignus; Chaitophorus* spp. Greenish to brown insects up to ⅛ inch long. Cluster on leaves, twigs, or branches. Suck sap.	Apply acephate, diazinon, or insecticidal soap.	
Stickiness and blackening of foliage from honeydew and sooty mold. Possibly some decline or dieback of woody parts.	SOFT (UNARMORED) SCALES: COTTONY MAPLE SCALE, *Pulvinaria innumerabilis;* BROWN SOFT SCALE, *Coccus hesperidum.* Immature scales are yellow to brown, immobile, flattened, oval insects less than 1/16 inch long. Adult cottony maple scale looks like popcorn. Adult brown soft scale remains flattened and brown. Suck sap.	Normally held in check by natural enemies. If not, apply carbaryl or diazinon whenever crawlers are numerous.	Cottony maple scale has 1 generation each year. Brown soft scale has 3 to 5 overlapping generations each year.
Decline or dieback of twigs or branches.	ARMORED SCALES, including: OYSTERSHELL SCALE, *Lepidosaphes ulmi,* GREEDY SCALE, *Hemiberlesia rapax;* SAN JOSE SCALE, *Quadraspidiotus perniciosus.* Immobile, grayish to brownish encrustations on twigs and branches. Individual scales are less than 1/16 inch long and circular, oval, or elongate in shape. Suck sap, may inject toxic saliva into plant.	Various natural enemies normally prevent these scales from reaching damaging levels. If not, apply oil during dormant season. Or, apply diazinon when crawlers are numerous in spring and summer.	A heavy scale infestation is not likely to be reduced to low levels by a single oil application. Several consecutive years of annual oil sprays likely will be needed. Oystershell scale has 2 generations each year in southern California; 1 in northern California. Other scales have several generations each year.

(continued)

Table 7. Willow *continued*

Symptoms/signs	Pest and description	Management options	Remarks
Decline or dieback of branches or sometimes entire tree.	BORERS, including: FLATHEADED BORERS (BUPRESTIDAE); ROUNDHEADED BORERS (CERAMBYCIDAE); CLEARWING MOTHS (SESIIDAE); CARPENTERWORM, *Prionoxystus robiniae*. Larvae are off-white, from ¾ to 2½ inches long when mature, and mine beneath bark or in heartwood.	Keep trees in good state of vigor. Cut off and destroy dying branches whenever seen. To kill established carpenterworm or clearwing moth larvae, inject ethylene dichloride into entrance holes, then plug entrance.	Most borers attack declining trees or tree parts, and hasten the death of the tree or its branches.
Skeletonized leaves.	LEAF BEETLES, including: *Altica bimarginata; Chrysomela aeneicollis, C. scripta;* CALIFORNIA WILLOW BEETLE, *Melasomida californica; Syneta albida*. Adults are brown to metallic black, oval beetles, ¼ to ⅜ inch long. Larvae are dark in color and elongate.	Apply carbaryl.	
Chewed leaves. May be silken tents in tree.	CATERPILLARS, including: SPINY ELM CATERPILLAR, *Nymphalis antiopa;* FALL WEBWORM, *Hyphantria cunea;* RED-HUMPED CATERPILLAR, *Schizura concinna;* WESTERN TUSSOCK MOTH, *Orgyia vetusta*. Naked, spiny, or hairy larvae, ¾ to 2 inches long when mature.	If caterpillars are confined to one or just a few branch terminals, prune off to remove colony. If not, apply *Bacillus thuringiensis*, carbaryl, or acephate.	
Prominent swellings (galls) on leaves that often turn red. May be globular or elongate and up to ½ inch long.	WILLOW LEAF GALL SAWFLY, *Pontania californica*. Insects responsible for galls are not often seen.	Galls are not known to injure the tree.	

Wisteria

Symptoms/signs	Pest and description	Management options	Remarks
White encrustations on woody parts and leaves.	WISTERIA SCALE, *Chionaspis wistariae*. Individual scales are elongate and less than 1/16 inch long. Suck sap.		Effect of scale on plant is unknown.

(continued)

Table 7. Wisteria *continued*

Symptoms/signs	Pest and description	Management options	Remarks
Decline of plant.	SPOTTED TREE BORER, *Synaphaeta guexi.* Larvae (roundheaded borers) are off-white legless grubs about ¾ inch long at maturity. Tunnel in woody parts.	Management never investigated.	Attacks injured and dying wisteria.

Yucca

Symptoms/signs	Pest and description	Management options	Remarks
Decline and dieback of plant parts.	ARMORED SCALES: OYSTERSHELL SCALE, *Lepidosaphes ulmi;* OLEANDER SCALE, *Aspidiotus nerii.* Immobile, grayish to brownish encrustations on plant. Individual scales are less than 1/16 inch long and circular to elongate in shape. Suck sap, may inject toxic saliva into plant.	Various natural enemies normally prevent these scales from reaching damaging levels. If not, apply diazinon when crawlers are numerous in spring and summer.	Oystershell scale has 2 generations each year in southern California; 1 in northern California. Oleander scale has several generations each year.
Blackening of plant from sooty mold. Cottony waxy material on plant.	LARGE YUCCA MEALYBUG, *Puto yuccae.* Powdery white oval insect about ⅛ inch long when mature. Suck sap.	Normally under biological control by natural enemies. If not, apply diazinon or malathion.	
Decline of plant. Holes punctured in leaves.	YUCCA WEEVIL, *Scyphophorus yuccae.* Adult is black snout beetle over ½ inch long. Larvae are white legless grubs that tunnel in base of green flower stalks and heart of plant.	Management never investigated.	

References

The first five publications are out-of-print, but may be found in agricultural college or university libraries.

BROWN, L. R., and C. O. EADS

1965 A technical study of insects affecting the oak tree in southern California. *California Agricultural Experiment Station Bulletin 810.* 105 pp.

1965 A technical study of insects affecting the sycamore tree in southern California. *California Agricultural Experiment Station Bulletin 818.* 38 pp.

1966 A technical study of insects affecting the elm tree in southern California. *California Agricultural Experiment Station Bulletin 821.* 23 pp.

1967 Insects affecting ornamental conifers in southern California. *California Agricultural Experiment Station Bulletin 834.* 72 pp.

ESSIG, E. O.

1958 *Insects and mites of western North America.* Macmillan Co., N.Y.

Farm Chemicals Handbook

1987 Meister Publishing Co., 37841 Euclid Ave., Willoughby, OH 44094.

FURNISS, R. L., and V. M. CAROLIN

1977 *Western forest insects.* U.S. Department of Agriculture, Forest Service Miscellaneous Publication 1339. 654 pp. Superintendent of Documents, U.S. Government Printing Office, Washington, DC 20402.

The next two publications contain color photographs of many of the natural enemies that attack landscape pests, and are available from Division of Agriculture and Natural Resources Publications, University of California, Oakland, CA 94608-1239

IPM MANUAL GROUP

1982 *Integrated Pest Management for Walnuts.* Publication 3270. University of California, Division of Agriculture and Natural Resources. 96 pp.

1984 *Integrated Pest Management for Citrus.* Publication 3303. University of California, Division of Agriculture and Natural Resources. 144 pp.

JOHNSON, W. T., and H. H. LYON

1976 Insects that feed on trees and shrubs. Cornell University Press, Ithaca, NY. 464 pp. Cornell University Press, 124 Roberts Place, Ithaca, NY 14850.

JOHNSON, W. T.

1985 Horticultural Oils. *Journal of Environmental Horticulture* 3(4):188-91.